Ranch in the Slocan

A Biography of a Kootenay Farm, 1896–2017

COLE HARRIS

1 2 3 4 5 — 22 21 20 19 18

Harbour Publishing Co. Ltd.
P.O. Box 219, Madeira Park, BC, VON 2HO
www.harbourpublishing.com

All cover photos are from the author's collection
Edited by Pam Robertson
Indexed by Joanna Bell
Cover design by Diane Robertson
Text design by Shed Simas / Onça Design
Printed and bound in Canada
Printed on acid-free paper certified by the Forest Stewardship Council and the Rainforest Alliance Council

Harbour Publishing Co. Ltd. acknowledges the support of the Canada Council for the Arts, which last year invested $153 million to bring the arts to Canadians throughout the country. We also gratefully acknowledge financial support from the Government of Canada and from the Province of British Columbia through the BC Arts Council and the Book Publishing Tax Credit.

Library and Archives Canada Cataloguing in Publication

Harris, R. Cole (Richard Cole), 1936-, author
Ranch in the Slocan : a biography of a Kootenay farm, 1896-2017 / Cole Harris.

Includes index.
Issued in print and electronic formats.
ISBN 978-1-55017-823-4 (softcover).--ISBN 978-1-55017-824-1 (HTML)

1. Harris, Joseph Colebrook. 2. Ranchers--British Columbia--Slocan River Valley--Biography. 3. Bosun Ranch (B.C.)--History. 4. Slocan River Valley (B.C.)--History. 5. Biographies. I. Title.

FC3845.S595Z49 2018 971.1'62 C2018-900589-0 C2018-900590-4

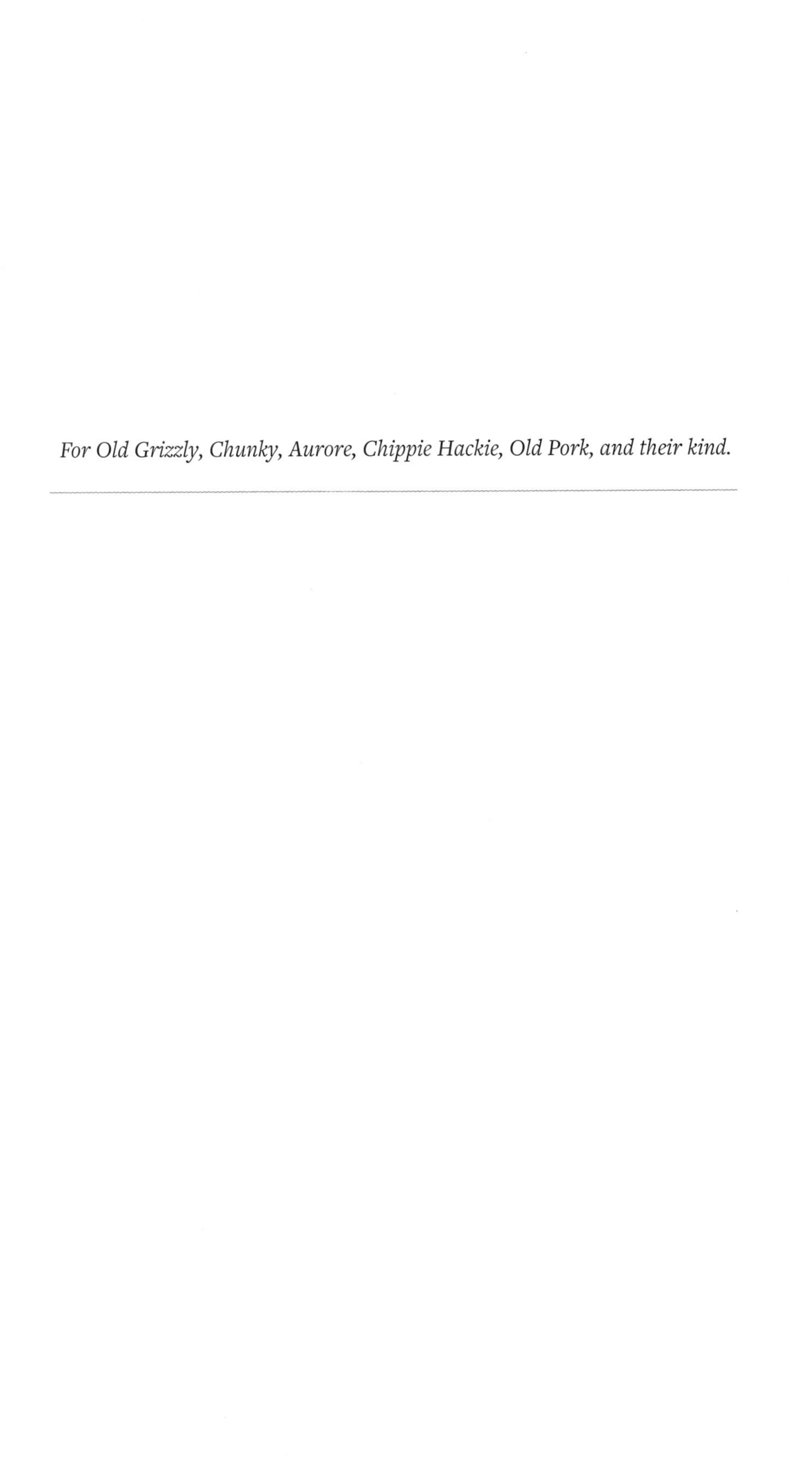

For Old Grizzly, Chunky, Aurore, Chippie Hackie, Old Pork, and their kind.

CONTENTS

INTRODUCTION

AT SOME TIME DURING THE LAST ICE AGE, A GLACIER FILLED THE LOWER reaches of British Columbia's Slocan Valley. Water pouring off this glacier and the surrounding mountains created narrow, milky, silt-laden lakes between the ice margin and the mountainsides. Each summer there were new loads of silt. Year by year, as the silt settled, the lakes filled with beds of clay. When the glacier advanced a little, the clay beds were covered with ice, and when it retreated, it left till and boulders on the clay. Eventually the valley glacier disappeared entirely, and the deposits of clay that had formed along its margin became narrow terraces above Slocan Lake. Over the years, largely coniferous forests—red cedar, hemlock, Douglas fir, white pine, and larch, in shifting proportions due to fires, forest successions, and climatic change—colonized the terraces.

At least four thousand years ago, perhaps much earlier, small bands of hunting, fishing, and gathering peoples moved into the valley. Eventually there were pit house villages along the Slocan River, a principal source of migrating salmon. People also fished in Slocan Lake and some of the creeks flowing into it, and ranged throughout the valley, including the proglacial terraces, to hunt and gather. The pre-contact population of the valley cannot be known and certainly fluctuated, but there is reason to think that it was often substantial. However, Old World diseases—smallpox, measles, and influenza—began to arrive, smallpox as early as the 1780s; coupled after 1826 with the attractions of a fur trade post (Fort Colvile) on the Columbia River south of the present international border, they drained the Slocan Valley of most of its people. Late in the nineteenth century, when Europeans began to

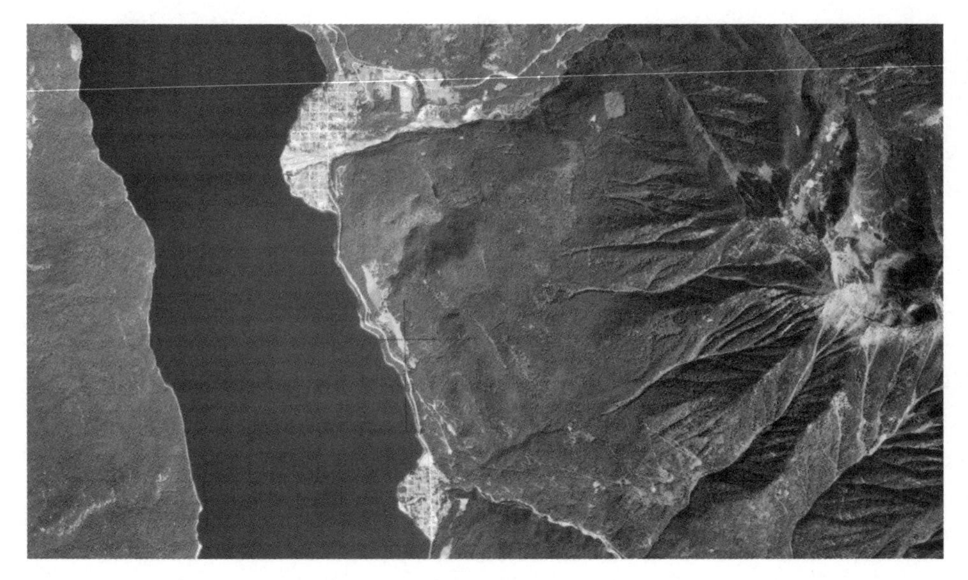

The slip of cleared land alongside Slocan Lake, midway between the delta villages of New Denver (top) and Silverton (bottom), comprises the fields of the Bosun Ranch. Province of British Columbia, 1982, 15BC82032 No. 180

arrive, the Indigenous population in the valley was far smaller than it had been for many centuries.

In 1888, a prosperous industrial family in Calne, Wiltshire, sent one of its younger sons, a lad judged to have no head for business, to Ontario Agricultural College in Guelph to learn to be a farmer. He left behind a large house stuffed with high Victorian comfort and servants; a family of successful businessmen, Gladstonian Liberals, and evangelical Christians; and no end of political talk and moral fervour coupled with a strong sense of the immanence of God. In 1896, this young Englishman—my grandfather, Joseph Colebrook Harris—fetched up on one of the terraces overlooking Slocan Lake. He cleared land, established a farm, and lived on it for the rest of his life. Calne and its larger English ways had converged with a mountainside terrace in British Columbia. My family has been involved with the results of this meeting ever since.

The chapters that follow are a set of glimpses of my grandfather's farm, which became known as the Harris or Bosun Ranch, and the people who lived on it at various times over the last 120 years. The past leaves a fragmentary record, and although many in my family were writers and many of their writings survive, their leavings are patchy. So, therefore, is this account. I have not tried to fill the gaps. The glimpses that follow depend on troves of surviving data: letters, diaries, my grandfather's writings about the early days, notes from conversations years ago, oft-repeated family stories, collections of family photographs, my own recollections, and in a few cases, published reports and official records. It is as if I, a scavenging bird flying over this unlikely farm, swooped down here and there where the pickings seemed promising. Overall, I have wanted to suggest how, on this mountainside patch in southeastern British Columbia, numerous lives and a strip of proglacial land have interacted over the last 120 years. In its details this is a very particular story, but elements of it have repeated themselves across Canada. This country's settlers have lived with bounded space, of which the Bosun Ranch is an extreme example, and have dealt with cultural change, as ways of life worked out in one place have been recast when re-contextualized in another.

I begin with Calne, where my grandfather grew up, then follow him to Guelph, Saltspring Island, the Cowichan Valley, and eventually the Slocan Valley (chapter 1). I use his own reminiscences to describe his search for agricultural land in the Slocan, his selection of the terrace that became the Bosun Ranch, and his first long year of work there (chapter 2). Information about his marriage to my Scottish grandmother and the development of the ranch through its first two decades is sparse, but I use the available fragments to describe these crucial ranch years as best I can (chapter 3). Using primarily my grandfather's account, I describe the discovery of a silver-lead mine on his property, then follow the uncertain career of the Bosun Mine—of industrial capital—in the midst of a Kootenay farm (chapter 4). In 1924 my father, then a graduate student in English at McGill University, returned to the ranch; his diary of that year provides a glimpse of the ranch's changing economy and of its political soul (chapter 5). I then turn to the project for radical social change that was at the heart of my grandfather's life, tracing it to its Fabian roots (chapter 6). A set of pictures taken from my mother's photo album allows me to describe the exquisite log cabin she and my father built in the 1930s (chapter 7).

This brings me to World War II. To this point, the Bosun had been my grandfather's ranch, but in 1942 the federal government took it over for several years, turning it into a camp for interned Japanese Canadians. I write about the family's, and to a degree New Denver's, relations with the internees (chapter 8). In 1951 my grandfather died, leaving a will that roiled the family for years (chapter 9). Eventually, most of the ranch became my Uncle Sandy's. He lived there all his life, loved its wild creatures, and used it as never before (chapter 10). In the early 1970s, restoration of a corner of the ranch house brought us in touch with the American counterculture that had recently poured into the Slocan Valley. I describe this convergence, the remarkable young Americans who remade the house, and the result of their work (chapter 11). In the 1980s, the second Bosun generation yielded, not without tension over different ways of being on the ranch, to the third (chapter 12). And a few years ago, a group of experienced Canadian back-to-the-landers—critics, like the hippies, of urban industrial ways but far more able to live comfortably at the edge of the bush—built a light clay house largely from local materials at an inmost edge of the terrace and just beyond the barn (chapter 13).

All the while, the fate of the whole ranch hung in the air. My cousin Nancy and her husband John owned most of it, but had no children. It was not clear whether the ranch would become a nature conservancy, be sold out of the family, or be turned into a high-end real estate development. My side of the family recoiled at some of the options. Now, the fate of the ranch for at least the next generation or two has been settled. In an epilogue, I explain what has happened, and muse a little about the past and future of the Bosun Ranch.

Cole Harris
Vancouver, BC
February 2017

1 CALNE TO COWICHAN

IN THE LAST YEARS OF THE NINETEENTH CENTURY, THE TERRACE ABOVE Slocan Lake that would become the Bosun Ranch attracted a prosperous young Englishman and a good deal of British money, a tiny edge of a vast outpouring of educated people and capital during the late heyday of the British empire. It was a bare moment in time, but long enough during the decades before World War I to establish the surplus young of comfortable British families in many corners of the world. Within this general exodus, the particulars were always different. In the case of the Bosun Ranch, they led back to Calne, a small town in Wiltshire, England.

In Calne, as elsewhere in the British Isles, the advent of the industrial age opened a route to wealth and power independent of landed property, commercial capital, or positions in government or the Church. The men who developed the new equipment and built the new factories tended to come out of the trades, and relied on practical ingenuity and business skills for their success. Such was the case of the Harris family in Calne. The first about whom there is any information, John Harris (1760–91), was a modest pork butcher. Out of this beginning the family built a pork-curing business that, before the end of the nineteenth century, processed more than three thousand pigs a week, supplied the royal family with hams, bacon, and sausages, and shipped worldwide. The Harris factories became by far Calne's principal employers; for years one Harris or another was usually the mayor. In less than a century, the family rose from obscurity to prominence, and it was from this background of new wealth and privilege that, in the fall of 1888, Joseph "Joe" Colebrook

Harris, my grandfather, was sent to the Ontario Agricultural College in Guelph to learn to be a farmer.

He left Guelph a year later, crossed the continent on the Canadian Pacific Railway, and settled in British Columbia, first in the Cowichan Valley on Vancouver Island, later in the Slocan. The following chapters turn around the farm he made above Slocan Lake, and the people who lived and worked there. But he came from Calne, and I begin there.

IN CALNE, THE HARRIS FAMILY TURNED A LOCATION ON THE ROUTE BY WHICH Irish pigs were shipped from Cork, Ireland, to Bristol, England, and driven to the London market, into the principal point where they were slaughtered and processed. How they did it is not entirely clear, but by the 1830s and early 1840s, as their business enlarged and shifted from butchering to curing, grandsons of the original John Harris travelled to Ireland to secure business contacts, and Harris bacon became known in London. When the Irish famine struck and the supply of Irish pigs dried up, one of the grandsons, George, then only twenty-two, visited the United States on an uncle's credit and with letters of introduction, intending to buy cured bacon and ship it to England. Ranging as far west as St. Louis, Missouri, he met the most important American meat packers, and visited several ice houses used for curing bacon in hot weather. The Harrises had cured bacon in winter, then hard-salted it to last through the summer. The intended advantage of the ice house, with ice on iron floors above and curing rooms below, was a mild-cured bacon available all year round. For a decade after George's return, the family experimented with various ice house designs, and eventually patented one of them: a large, thatched, barn-like structure built of iron and stone with charcoal between double walls, capable of holding up to one thousand tons of ice.

In the late 1880s the Harrises adopted a mechanical and chemical process for cooling brine, then used as a coolant. The ice houses became redundant, but during their time they apparently had been the Harrises' main technical advantage. Thomas Harris, my great-grandfather, thought that his younger brother George, who brought the idea of the ice houses

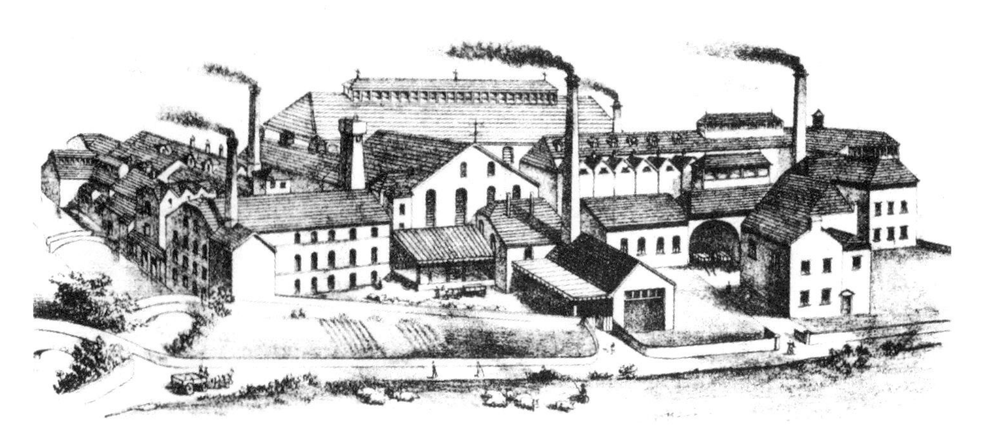

The Harris factories in Calne around 1888, the year Joe was sent to Ontario Agricultural College. The large, hip-roofed and thatched structure in the background is an ice house. This image is taken from a brochure for the exhibition *A Taste of Harris* at Marden House, Calne, November 1992, prepared by the Heritage Centre Working Group of the Calne Town Council.

from America and died at the early age of forty-five, "was the smartest business man of any of us; he was the means of lifting us out of the old rut, and he laid the foundation of the new system and its prosperous future."[1] For many years there were two Harris firms that between them were the principal suppliers of pork products in the British Isles. In 1885 my great-grandfather, the owner of one firm and getting on in years, brought his three eldest sons into the business: Thomas Harris & Sons.[2] Joe was then fourteen years old, too young to be considered for the partnership, but probably already sized up as an unlikely businessman.

In 1865 Thomas bought a three-storey, eighteenth-century house at a prominent Calne intersection, enlarged and embellished it, and named his acquisition "South Place." From the street, South Place was an awkward mass linked at a central corner by bay windows and a parapet; from the walled gardens at the back, a much more elaborate, somewhat Italianate, mid-Victorian creation with its own fernery and grotto. South Place had nothing to do with landed gentry. It was a prosperous businessman's urban residence, quite large enough for a numerous family, their

South Place, Calne, no date (n.d.) but probably 1880s. The original building, on the right, dates from the eighteenth century. The rest is Victorian. Note the balustrade, the machicolated cornice, and the bay windows. South Place fronted directly onto the street; its elaborate gardens were behind. Harris family photo

servants, and events for company employees. Joe and his many surviving siblings—four by my great-grandfather's second marriage, four by his third—grew up there.

Thomas Harris's family belonged from 1866 to what was known as the Free Church. The local Anglican parish had been served for years by an evangelical minister, but when he died and the Bishop of Salisbury replaced him with a High Church minister, many of the congregation, the Harrises prominently among them, strongly opposed him and eventually established their own evangelical church. Evangelical Christianity was in the South Place air. Morning and night, family and servants assembled for prayers; Sunday was given to church and little else, and cards and alcohol were always proscribed. Thomas was president of the West of England Temperance Society. The family gave heavily to the evangelical London Missionary Society, and two of its daughters went to China as missionaries. Deaths, even of the very young, were somehow God's will, and heaven was beyond. However inscrutable, God's will was always done, always for the best.

Terraces, balustrades, and gardens behind South Place, n.d. but probably 1890s. Two full-time gardeners maintained the gardens. Harris family photo

One senses a particular religious fervour when Thomas's third wife, my great-grandmother, arrived in 1866. My father said that his grandfather married first (Susan Reynolds) for love, second (Sophia Mitchell) for business, and third (Elizabeth Colebrook) for religion. Elizabeth came from a well-known Nonconformist family near Guildford in Surrey. Her father, a businessman and farmer, was also a lay preacher esteemed for his devotion to Calvinist principles and commitment to social service, which apparently ran deep in the extended Colebrook family.[3]

Thomas Harris was a Liberal, and politics a substantial part of his life. He never ran for national office, but vigorously endorsed Liberal candidates, contributed financially to their campaigns, and often chaired political meetings. Economically, he believed in free trade and open markets; politically he supported universal suffrage (for men), the elimination of rotten boroughs, and home rule for Ireland; and socially he supported measures to alleviate the lot of the poor. Prosperous as he was, poverty was not many generations behind. At the end of his life, it was said of him that he opposed "everything that savoured of oppression and intolerance."[4] He was a staunch supporter of William Gladstone, even when the country turned against his plans for Irish home rule and many members of the extended family joined the Conservatives. He was elected five times as mayor of Calne.

Encouraged one suspects by his third wife, Thomas attended innumerable peace meetings, prayer meetings, and temperance meetings, and gave a great deal of money to a variety of causes. "Enos Gale our old coachman," Joe remembered years later, "was forever driving father with a party of supporters to public meetings until my Dad became known as The County Chairman."[5] He supported the local Free Church, small Nonconformist chapels elsewhere in Britain, the London Missionary Society, and sanitariums for convalescing missionaries in China. He supported temperance organizations. He bought and converted a public house into a coffee house. He poured money into Calne: almost a quarter of the cost of a new city hall, a public recreation ground and pavilion, a reading library for working men, money for the poor to be distributed at the discretion of the town council. He provided in various ways for his employees, apparently paying them fairly well and organizing annual excursions, dinners, and gifts: a Bible to each of them after one company dinner, clothing for needy families at Christmas. In an after-dinner speech to employees, friends, and prominent townspeople he is reported to have said that he had grown up from boyhood with some of his workmen, hoped to be their master for a good time yet, and considered that the bond between master and men would not be merely mercenary as long as the interests of both were promoted.[6]

Life at South Place, the family seat in Calne, coupled reformist zeal and business achievement. "Our home," Joe remembered years later, "was large and very comfortable indeed, with the most solid and British type of comfort. It was jammed into the little town of Calne with its fair-sized and wonderfully-well-kept garden. Calne was full of Harrises, some quite rich and all in comfortable circumstances. We had an amazing number of visitors; it [South Place] was like a hotel. Most of the speakers on peace, temperance, religion, and liberalism who came to Calne seemed to find their headquarters at South Place. Mother was more especially interested in religious matters and especially missionaries, so we saw very many most interesting and often amusing people."[7]

Joe and his siblings were all sent away to school: the two girls (Bessie and Mary) to Miss Fletcher's School, an outstanding private school for girls; the boys (Willie, Joe, and Alec) to Mill Hill, a Nonconformist school in London—both carefully chosen to avoid the danger of Darwinian contamination. At Miss Fletcher's the atmosphere was relentlessly Christian, and missionaries and missionary work greatly esteemed. Mill Hill was expensive and flourishing, a school where the sons of the newly rich were intended to be rendered into gentlemen. Suspicious that the new science was undermining religion, Thomas carefully inspected the school and met one of its masters, Dr. John Murray, who later became the editor of the *Oxford English Dictionary*. Murray was a Liberal and a temperance man; apparently the two got along and Thomas was reassured. At Mill Hill, Willie and Alec became impressive young scholars and Joe excelled at cricket and rugby.

After Mill Hill the question of what to do, common to the numerous progeny of prosperous English families, was suddenly in the air. Joe was a strong, athletic young Englishman, and an outstanding rugby player,[8] but he had no obvious vocation. Three sons by his father's second marriage had filled the available positions in the family business, for which, anyway, Joe had neither aptitude nor interest. Willie and Alec were headed to Cambridge. In these circumstances, Thomas turned to the empire. He heard from a friend about the Ontario Agricultural College in Guelph, Ontario, made extensive enquiries, and decided to send his son there. Joe, born in January 1871, was eighteen years old. Like the other young Englishmen at the Ontario Agricultural College, he knew absolutely nothing about farming in Canada.

AFTER MILL HILL AND LONDON, JOE FOUND THE ONTARIO AGRICULTURAL College rough and ready and Guelph drab but friendly. His fellow students at the college were young, transplanted Englishmen—most from private schools and many with aristocratic pedigrees—or solid Ontario farm boys. Most of the former, he wrote years later, were good at games, concerts, and play, but "few of us had the slightest notion when we arrived what farming in Canada was like and imagined ourselves riding about on dashing steeds and shooting game in the company of Indians and cowboys." Many of the young aristocrats he considered foppish, and one, a nephew of the Duke of Norfolk, among "the slickest blackguards and good-looking highly-trained genteel scamps I have ever met." The Canadians were "jolly good fellows," if, for the most part, unmannered and from uneducated homes. Compared to the English, however, they knew far more about "the real business of life."[9]

Joe's academic record at the agricultural college was mediocre, but as captain of the rugby team and editor of the OAC *Review*, he became a fairly prominent student. He seems to have made friends easily, principally among them Robert Musgrave, who had come to Guelph from a large sheep farm on Saltspring Island, near Victoria, British Columbia. The Musgraves were an old, titled Cumberland family with a branch in Ireland to which Robert's people were attached. He and Joe became fast friends, sharing a fondness for games and sport and a disinclination for study. Joe thought that his friend Robert had "a wonderful lot of experience in life," whereas his own life in Calne had been too sheltered. If so, a solution was apparently at hand. Robert invited Joe to spend the summer holidays with his family on Saltspring Island, and gave him explicit instructions about getting there. As soon as Joe's year-end examinations were over in 1889, he set out from Toronto, travelling in a colonist car full of central European peasants on the recently completed transcontinental line of the CPR. He had brought no bedding, and slept as he could on a plank upper berth.

The trip west was full of entirely new sights and peoples, and the destination a veritable paradise. Joe took a coastal steamer, the SS *Joan*, from Victoria, called in at "strange little ports and landings," saw "Indians" in canoes and on the wharves, "Chinamen" in pigtails, whales "spouting and jumping almost like trout," hair seals, "eagles circling around and osprey diving for fish," and eventually arrived at Musgrave Landing on Saltspring Island.[10] Before the day was out, he was playing tennis on the Musgraves' wooden tennis court, then deer hunting.

The Musgraves considered it improper that a guest work, and with the exception of a few days during the sheep drive (the Musgraves ran 1,200 Merino sheep), Joe's summer with them was an extended holiday.

He and the Musgrave boys fished, hunted, swam, and played tennis. Jim, "the house Chinaman," did much of the housework; Lum, the outdoors man, did much of the garden and yard work. Joe did take on a large, somewhat rotten cedar, and after hacking at it with a dull axe for hours managed, amid cheers from the Musgraves, to fell it. He also helped for a couple of weeks with haying on a farm belonging to the family of another Guelph classmate, Pat Johnson, at Hall's Crossing (later called Westholme) in the Cowichan Valley on Vancouver Island. There he got to know Captain C. E. Barkley, RN, prominent among the retired Royal Navy officers in the Cowichan, "a real old sea dog and a tremendous talker ... only always on the Conservative side."[11] Essentially, this was a summer of play. The Musgraves had a good rowboat and a fishing smack (the *Jabberwock*) with an auxiliary engine, and on Saturday mornings used one or the other to get to the tennis courts at Cowichan Flats. Tennis was the settlers' principal sport, and the weekly convergence at Cowichan Flats a major social event. There was also a midsummer tennis tournament in Victoria, to which the Musgraves repaired in the *Jabberwock*. Joe and the Musgrave boys stayed with the Crease boys in Judge Crease's large, Italianate house, Pentrelew, on Fort Street. Crease was a former attorney general and a justice in the BC Supreme Court; his wife, the daughter of an English literary critic who was a friend of Charles Dickens, had often sat on Dickens's knee while he and her father talked. Tennis, dinner parties, dances in Victoria—an expatriate, upper-class English life that readily accepted Joe and into which he easily fit.

He greatly admired Mrs. Musgrave, and was pleased that, in spite of his politics, she rather took to him. Her husband was more distant and less approachable—"rather awesome, quite the old aristocrat"—and judged to be ruled by his liver. The Anglo-Irish Musgraves were staunch Conservatives: Irish home rule anathema, Gladstone an evil monster. In such company, Joe apparently held to his opinions. He got on well with the young Musgraves, liked the two Chinese servants, Jim and Lum, and liked what he saw of the local Indigenous people. When they came to assist with the sheep drive, they were "a picturesque sight" around their campfires and seemed friendly and full of fun. He wished he could understand their jokes. He said that the settlers despised them, considering them lazy and degraded, but noted that when he got up at three a.m. to fish they were always on the fishing banks before him.

Overall, and for all his fondness for the Musgraves and delight in a summer that reshaped the course of his life, his analysis of settler society was critical. "It seemed to me that the outlook on life was extremely narrow and intensely conservative. They [the Musgraves and their Victoria friends] tended to despise Eastern Canadians and Americans and had a little circle of similarly minded friends in Victoria who had become completely fossilized without the slightest idea of their condition. It was a great contrast to the sentiments and mental outlook of South Place, Calne."[12] British Columbia's abundant nature and exotic mix of peoples had captivated my grandfather, not the transplanted English society that had treated him so well.

After such a summer, Guelph was a huge letdown. College life was dull, and English students were often blamed for playing poker and drinking in the smoking room. In these circumstances, my grandfather wrote to Pat Johnson in Westholme, whose family he'd helped with haying, to ask for a job, received an encouraging reply, and announced that he was leaving. When asked why, he said that Englishmen appeared to be unwelcome at OAC, a response that caused much consternation at the college and much anxiety in Calne, and that eventually he much regretted. Basically, he longed to return to BC. When he left in early December, the college president wished him well and the college matron packed him a basket of food. After buying another colonist car ticket and, this time, a cheap straw mattress for an upper berth, he was almost broke.

On the Johnson farm in Westholme, he found himself alone with Pat, who was by then working the farm for his father. He and Pat batched in a shack set on cedar blocks. Pat was a worker, as was Joe, but they do not seem to have gotten along. Nor was there much connection with neighbours, whom the Johnsons and the Barkleys judged inferior. Pat left for Victoria at Christmas, leaving Joe to reflect on Christmases at South Place, where "far too much was done for us." Then Captain Barkley, also alone (save for his Chinese cook), came over to invite him to Christmas dinner. Joe liked Captain Barkley, "a fine old fellow," and accepted with great pleasure. Although politically they disagreed "profoundly but politely," he long remembered this Christmas dinner, served for two by a Chinese cook at an edge of the emerging settler landscape in British Columbia, "a glorious evening, just we two."[13]

Joe remained on the Johnson farm well into the new year, and in so doing got to know another English family, the Gibbs, who lived some four miles away. The Gibbs were impecunious gentry. A Gibb grandfather had been "a 'Queen's Messenger,' some sort of very confidential royal servant and they had not got over that unfortunate start, for Queen's messengers were rather out of place on a Vancouver Island stump farm." When Joe first met the Gibbs, two of the sons "and an old fellow named the Bosun," who took on odd jobs in the area, were clearing land with a stumping machine. Joe lent a hand, thoroughly enjoyed himself, and stayed for supper. "Bosun baked a fine big batch of biscuits ... Afterwards we had a great sing song. Old Bosun knew very many good old sea songs and was a very good singer. I knew all the Guelph songs, mostly quite new to the Gibbs."[14]

Sometime in the spring of 1891 it was determined in Calne that Alec, Joe's younger brother, would visit his brother in British Columbia during the Cambridge summer holidays, and they would tour America together before both would return to England. Perhaps it was a rescue mission. With his brother's arrival in mind, Joe left the Johnson farm and went over to Saltspring Island for a few days to visit the Musgraves, then to Victoria to visit the Creases. He had a friend in Chilliwack, one of the Wells family, who had also attended OAC and had told Joe about the wonders of the Fraser Valley and of his family's farm and Ayrshire cattle. Joe therefore thought to go to Chilliwack, meet the Wells family and see something of the Fraser Valley before meeting his brother there. On the boat travelling upriver from New Westminster, he met an Anglican parson who invited him to get a horse and ride with him the next day to meet some of his parishioners, which he did and met a young English couple

The Musgrave house on Saltspring Island, 1891. As Alec wrote, "We spent a few lovely days there, doing nothing in particular—tennis, bathing, coon hunting, fishing and afternoon tea." From Alexander Harris and Joseph C. Harris, "Our Diary in America from June 17–Oct. 1, 1891." Copy in the Royal British Columbia Museum and Archives, original in Harris family papers

in need of farm help. He stopped with them for a long month, and saw little of the Wells family. Shortly thereafter, his brother arrived by train from Montreal.

The two brothers went over to Vancouver Island where they fished, hunted, walked—twenty-two miles to Cowichan Lake—and visited Joe's old haunts and friends. Alec had a Kodak camera, still unusual enough to attract attention, and took snap shots and pasted them in his diary: Macdonald's logging camp on Cowichan Lake, the hay wagon on the Johnson farm, the Musgrave house on Saltspring Island, Musgrave children and Joe on the front steps, and, a mirror of transplanted English gentility on Vancouver Island late in the nineteenth century, afternoon tea on the lawn. They decided to go "up country," crossed the strait to Vancouver, and took the train as far as Ashcroft. They arrived in the middle of the night, and woke up in what seemed a desert. Alec wrote in his diary: "nothing but sand was visible, sand everywhere, rising all around into lofty mountains of the same monotonous yellow, and the only growth was the wild musk plant, which looks in the distance like the sand itself."[15]

Afternoon tea on the Musgrave lawn, 1891. Mr. Musgrave is under the tree, right, Mrs. Musgrave is in the centre background, and a Miss Annis is under the white hat. Joe (or more probably Alec) is lying in front of the bush and behind Robert Musgrave. From Alexander Harris and Joseph C. Harris, "Our Diary in America from June 17–Oct. 1, 1891." Harris family papers

In a middle-of-the-night rush to get the train on their return, they left the Kodak behind. Telegrams did not retrieve it. They waited a couple of extra days in New Westminster, then gave up and carried on to the United States.

In the United States, their small caps and other English clothing identified them as what they were: two young, sightseeing Englishmen. From San Francisco, they walked much of the way into Yosemite Park, then went south to Monterey. They formed strong opinions: American towns all looked the same; American girls were great flirts; the almighty dollar governed American life. They thought Salt Lake City a pretty town, felt after seeing the Chicago stockyards that they never wanted to look another cow in the face, and were overwhelmed by Niagara Falls. Alec thought Guelph and the agricultural college much more attractive than his brother's letters had led him to expect. After Montreal, they went to New York and boarded a transatlantic liner, the *Teutonic*, for England. By the beginning of October 1891, they were back in Calne.

CALNE SEEMED MUCH AS JOE HAD LEFT IT, AND SOUTH PLACE "AS SOLID AND comfortable as ever." England, too, seemed "secure and prosperous ...

on the surface at least."[16] Other than pike fishing with the headmaster of the older boys' school, he had little to do in Calne, and soon went up to London to stay with his half-sister Bessie and brother Willie, both finishing their medical studies—Willie at Guy's Hospital and Bessie at the London School for Medical Women at the Royal Free Hospital. He joined the Rosslyn Park Football Club, the premier rugby club in North London, was soon on the first team, and played against some of the best teams in England, and also against his former school's team, the Old Millhillians. He took a course in English literature at University College, joined the University College Debating Society, becoming a leader of the radicals, and attended meetings of the Shakespeare Society. Overall, he was not impressed by University College: it seemed "a kind of knowledge shop, where everyone went to carry off lots of facts." Even in London, he had time on his hands, and seems to have spent much of it with wealthy friends, particularly a young heir of a Lancashire cotton fortune, who, until redirected by an inheritance, had been a medical student with Bessie and Willie. While a mansion was being renovated for him on Hampstead Heath, he hobnobbed around London with Joe, paying most of the bills. They heard a young David Lloyd George defend Welsh language rights in parliament, and played a lot of billiards.

In March 1892, Joe returned to Calne. "Everybody," he later wrote, was "very kind and friendly, but I could never fit in such stodgy surroundings ... I longed to get back to Canada."[17] He was now twenty-one years old, and, like his older siblings in their turn, had received a thousand pounds from his father on his birthday. With this considerable sum in hand, he went directly to Victoria, stopped there briefly, then went up-island to visit the Musgraves. "Newcombe Musgrave, who had been at school in England and who had visited us at South Place, travelled with me."

HE FOUND THE MUSGRAVES IN THE PROCESS OF MOVING FROM SALTSPRING Island to a farm near Duncan, and helped them with the move, then set about looking for land of his own. The process was not drawn out. The land across the swamp from the Johnsons' farm was for sale, and Joe bought it for a thousand dollars. The higher land was entirely and heavily forested, the lower a bog with, in the middle, a lake of some twelve acres. There was no access road, not even a trail. Between what he had bought and a farm was an enormous amount of work.

Joe threw himself, sons of local farmers, hired Chinese workers, and even the Bosun at this land. He and four Chinese labourers cut out a passable, half-mile wagon road. He and a neighbour, Jack Windsor, the son of an Englishman married to a Salish woman, tackled the forest. The cores of most of the huge cedars and firs were rotten, and by drilling two-inch auger holes from either side and using hot ashes to light fires inside the trees, they could burn them down. Often there was no telling where they would fall. On the ground, the trunks could be burned into lengths by the same method, then split with blasting powder. With a brace of oxen, the remains could usually be piled and burned. As in pioneer settlements across the span of North America, the forest was an enemy. As were the stumps. Joe tried dynamite, and considered it "wonderful that I did not blow myself up." Such work made the bog, which looked almost ready for ploughing if only the water could be got off, particularly attractive. A drainage ditch, at least a few feet deep and a good many hundred feet long, was needed. Joe dug the side ditch around part of the swamp, and contracted with Chinese workers for the main ditch. When both were finished, the bog on his side of the lake began to dry out.[18]

During the first summer, Joe lived at the Windsors' house, but left in the fall when Jack's father returned and went "on his usual big spree." By this time the Bosun had reappeared, and he and Joe batched for a time in a shack on an adjacent quarter section. The Bosun had spent thirty years in the Royal Navy until, while drunk one night, he struck an unpopular officer and was put in irons. The crew sympathized with the Bosun and managed to spring him. Somehow he got to Seattle and then to Vancouver Island. His alias was Charlie King; his real name, according to my grandfather, William Dyer, and his home the Isle of Wight. He seems to have floated throughout the Cowichan Valley, staying for a time here or there, but increasingly with Joe. He was a good companion except when he went off on a drinking binge from which, sheepishly, he would eventually return.

Another to appear at this time was Arthur Cleverley, a lad from Calne who had worked as a gardener, then a groom. Initially he seemed puny and, with his broad Wiltshire accent, completely out of place. "He had looked," wrote Joe, "such a mild and innocent youth when he arrived, but I soon found out that in comparison with him I was a mere babe in the woods." He had worked for many wealthy people (including, as Joe put it, his rich and "rather awful Aunt Charles"), was full of stories about their more egregious doings, and had a lightning wit and an ability to mime and caricature. He played the concertina and liked to dance; he liked the ladies, and they him. And he could cook. He became useful and very popular.

In sum, Joe had assembled an unlikely crew, all of whom he considered friends: a "half-breed" increasingly disparaged by white society, especially its women; a deserter from the Royal Navy; and a cheeky lad from Wiltshire. For a time, he, the Bosun, and Cleverley lived together in a room in the barn.

By the spring of 1894 Joe had a small kitchen garden and a little land that could be ploughed, which he planted in oats and red clover. He also arranged to have a house built—just in time, as it turned out, for a contingent of Harrises from Calne. While Joe had been establishing himself in the Cowichan Valley, two of his sisters, Bessie and Mary, had gone under the auspices of the London Missionary Society as missionaries to Hankow, China. While there, both had married other missionaries. Eighteen days after Mary's marriage, her husband died of amoebic dysentery, one instance among many of the carnage that disease and violence were wreaking among foreigners in China. In Calne, Thomas, the patriarch of the family, apparently decided to send out reinforcements, and also to give some of his children a trip around the world. Alec, who had finished a degree in engineering and physics at Cambridge, was available, as were Willie, who had completed a residency at Guy's Hospital, and Sophie, Joe's half-sister by Thomas's previous marriage. They were to meet up with

Joe on Vancouver Island and stay for a time, then Willie, whose fiancée was in London, would return to England, and the others would go on to China. After a visit there, Joe would take one of the Empress ships back to Vancouver Island, while the others continued around the world.[19]

This contingent descended upon Joe in September 1894, stayed for a month, and had a wonderful time. There was a spartan but fairly comfortable house, "a particularly attractive little Chinaman" called Golly to do the cooking, good hunting to occupy their days, and a surrounding society that Joe knew well and in the upper tiers of which they were entirely welcome. There were visits, teas, dinners, tennis, boating, and an abundance of nature that fit an English romantic imagination. Sophie, a fair artist, brought her oils and left Joe with two pictures of his new house: one amid ragged, burned-out stumps at the edge of a towering forest, the other from another angle and in a larger clearing, a space littered with the trunks of fallen trees and jagged stumps. The pile of logs in the foreground, inside the snake fence, is aflame. There is nothing yet like a farm.

The trip to China went ahead as planned. Joe was far more intrigued by the Chinese and Japanese people than by his sisters' missionary work.[20] After a couple of months in China, and long walks with Alec where they had been strongly advised not to go, on December 28 three Harris siblings departed in opposite directions from Shanghai: Sophie and Alec for India, Joe for Vancouver Island. Bessie and Mary remained in China where, six months later, Mary also contracted amoebic dysentery and died. Aboard the *Empress of India*, Joe finished *Principles of Political Economy* by John Stuart Mill (a gift from a Mill Hill friend) and found and devoured a well-worn copy of *Protection or Free Trade* by Henry George, a little book that "gave me my first real insight into the real problems of human relations." By the time he docked in Victoria, he felt himself a very different young man, and when he returned to the Cowichan Valley he ordered all of Henry George's writings. He had become, he said, a red-hot single tax supporter and a disappointment to many of his friends.[21]

As Joe settled back into the Cowichan Valley, their disappointment had reason to expand. A German neighbour, J. W. Hern, was a well-read

Joe's house and clearing, Cowichan Valley, 1894, oil painting by Sophie Harris. Note the many jagged burned-out stumps and a substantial new house, the product of English money rather than of any return from a farm barely begun. Harris family collection

and outspoken atheist and Marxist. He worked ostentatiously and noisily on Sundays, "was not properly respectful of his richer neighbours," and held, so almost everyone thought, "atrocious opinions about property." Joe liked Hern, considering him "wonderfully patient and good natured, a kind of ideal perverted Christian," and the two soon argued about religion and politics. Hern, who knew a good deal of contemporary German philosophy, apparently had much the better of it. "German philosophers and Herbert Spencer were altogether too much for my religious convictions, and I got to the state of denying God and laughing at the mythical story of Jesus and his miraculous birth ... Science was explaining many things and performing so many wonders, and no doubt it would proceed on its majestic way to settle all human problems. My poor former friends were greatly aroused and horrified by my fall from grace." That was not all. Hern lent Joe a copy of *Capital*, and after some struggle to familiarize himself with Marx's prose he was absorbed by this huge work, and became, he said, "a white hot socialist" bent on reforming the world. This while clearing land in the Cowichan Valley, "hardly the place," he noted, "to be influenced by radical opinions. There were very many highly respectable people there, living on remittances and pensions, who felt that Their World had little need of reform, and certainly none of reconstruction and revolution."[22]

In these ways Joe increasingly stood apart from Cowichan society, but in other ways he did not. He was captain of the local rugby team (which played amid stumps on a field in Duncan) and helped organize a theatrical society and performed in its first play, Sheridan's *The Rivals*. He admired Wilfrid Laurier who, he thought, had a vision of Canada that transcended the local, frequently anti-Canadian feeling common in the Cowichan Valley, and twice chaired political meetings in support of the federal Liberal candidate. And he actively supported a creamery, the first in British Columbia. But he was becoming restless, less because his religious and political opinions were seldom shared than because the agricultural economy of the Cowichan Valley was flat, there was little market for farm produce, and the outlook was not promising. People were leaving. The Musgraves sold their new farm and moved to Victoria; Robert Musgrave, Joe's friend from Guelph days, went to McGill to study engineering. "Even the real old 'hard work, pinch each nickel' farmers were having desperate times, and the young people were beginning to drift into the cities more and more." Joe, who thought the cities were filling up with useless people, did not want to join them, but what to do instead was a real question.[23]

It was during these uncertainties that his parents, who were getting old, asked him to come home. Sometime late in 1895 or early 1896 he went, sailing from New York. On shipboard he got to know a husky young fellow who turned out to be "Kid" McCoy, the world light heavyweight boxing champion. McCoy needed a sparring partner, and Joe, after insisting that he was a green amateur and urging McCoy to take it easy, did spar with him and apparently survived unscathed. After landing in Liverpool, he was soon back in the circle of the family: his aging parents; Sophie, who had become, virtually, a private secretary to her father; Willie, now a doctor in Shaftesbury; and Alec. Although "the crowd and company were not of my sort," he hunted hares, rabbits, and partridge with a large party of shooters and beaters on Salisbury Plain. And he joined the Fabian Society, the institutional home of intellectual English socialism. He told his parents a good deal about his farm in the Cowichan Valley and also about its depressed economic circumstances. They urged him to try New Zealand, where a Harris relative had married a Maori princess who controlled four thousand acres of prime timberland. It was a matter, apparently, of finding the right spot in the empire. However, with the beginnings of a farm already on his hands, Joe returned to Vancouver Island.[24] He would keep an open mind about the future.

Back in Westholme, he was again clearing and farm-making with the Bosun, Arthur Cleverley, and Jack Windsor amid the mixed society of the Cowichan Valley and a depressed economy. Sometime in the spring of 1896 a letter arrived from E. R. Pease, secretary of the Fabian Society, to inform him that two other Fabians, a Scot, Robert B. Kerr, and his English wife, Dora Forster, had moved to British Columbia and were living in New Denver, a new mining town on Slocan Lake in the Kootenay District of southeastern British Columbia. He suggested that Joe get in touch with them. He did; the Kerrs replied that they meant to settle in New Denver, but were going shortly to New Westminster, and accepted Joe's invitation to pay him a visit. They turned out to be an extraordinary pair of eccentric intellectuals. Kerr, a gold medallist from Edinburgh University and a lawyer by profession, was a close friend of Bernard Shaw and Beatrice and Sidney Webb, had a vast knowledge of English literature, and looked, Joe thought, like a youthful Mr. Pickwick. He was terribly shortsighted, and would "go slowly along, rubbing his hands, his wonderful eyes blazing with intellectual activity and his big feet rising and falling as if he was always anticipating some unseen obstacle." His wife wore an astonishing collection of unrelated colours. Dora seemed to be trying to be artistic, but in fact, Joe thought, "she just went her own way in great simplicity and awful untidiness." When she arrived at Westholme she wore a "crushed and unhappy" hat she had made of lichens from fir trees. Joe liked and admired the Kerrs, but his neighbours found them very strange. Cleverley and the Bosun were greatly amused.[25]

Learning from Joe of the depressed economy in the Cowichan Valley, this incongruous pair strongly advised him to consider the Slocan, then in the full spate of a mining boom. There was, they said, a great lack of fresh fruit and vegetables, which commanded famine prices, and an inexhaustible market. Joe promised to visit them in New Denver as soon as his hay was in and he found someone to mind the farm. In a few weeks he set off, accompanied as far as Revelstoke by one of the Crease boys. From Revelstoke he took the just-built CPR spur line to Arrowhead at the north end of the Arrow Lakes, then a lake steamer to Nakusp, "a most forlorn looking spot with two very rough hotels." From Nakusp, a dilapidated car, which seemed ready to leave the track at the slightest provocation, swayed and scraped against newly blasted rock walls along another new CPR branch line, the Nakusp and Slocan, and eventually deposited Joe at New Denver.

2 PROSPECTING FOR LAND

IN LATE 1943, SHORTLY AFTER THE DEATH OF HIS WIFE AND ON THE URGING of my mother, Joseph Colebrook Harris, my grandfather, began to write an account of his coming to Canada and eventually settling in the Slocan. Over the next several years it expanded into well over a hundred typed, single-spaced foolscap pages. His heart rate was very low and he was frequently dizzy and tired, but his long-term memory was sound. The principal defect of his account is that it ends in the fall of 1898. It does not include, therefore, his trip to England at the end of that year, his marriage and return to the Slocan with a wife, and the early years of their expanding family on what was intended to be a major commercial orchard. Yet there is a vivid account of his search for land in the Slocan, his selection of a relatively flat, well-watered, and densely forested terrace overlooking the lake, and the beginning of its agricultural development. I reproduce most of his account here,[1] leaving for chapter 4 his account of the discovery and sale of the prospects that became the Bosun Mine. Except for my few bracketed interventions, the words are his:

(As mentioned in the previous chapter, Joe came to the Slocan on the advice of Robert Kerr and Dora Forster, a pair of eccentric British intellectuals who lived in New Denver, visited Joe at his farm in the Cowichan Valley, and assured him that the opportunities for farming in the Slocan were boundless. So prompted, in late 1896 he came to see for himself, was enthralled by his first glimpse of Slocan Lake, and arrived at New Denver Siding on the CPR's newly completed Nakusp and Slocan Railway.)

Scene of the head of Slocan Lake, ca. 1900. As Joe wrote on first seeing it, "Then I caught my first glimpse of Slocan Lake, the very perfection of wild mountain scenery... I was certainly thrilled then, and still am." Silvery Slocan Museum, New Denver, Image 2001-002-038

KERR WAS AWAY when I walked down from New Denver Siding and was directed to their tiny house. He was at Revelstoke, where I might have met him on my way in. Mrs. Kerr gave me a most hearty welcome and insisted that I must stop with them until Robert returned in a day or two. I should have to sleep on the floor under the shelf in their tiny pantry. I was young and made the best of it, but it was, I should judge, at least six inches too short.

Mrs. Kerr was an enthusiastic gardener and longed to have a good flower bed right in front of the house. I agreed to try to dig the ground, but my heart sank when I went to look at the place, for it seemed one huge rock pile. New Denver lies on the delta of Carpenter Creek and Carpenter Creek is a fierce mountain torrent that has been busy for thousands of years carrying big boulders and gravel every spring freshet to extend her delta into the lake.

I hunted up the Kerr outfit of tools and concluded that lawyers do not know much necessarily about axes, picks and shovels.

New Denver, 1897. This early photo shows the denuded townsite at which Joe arrived in 1896. The forest was cut, and the land strewn with stumps and cobbles. Postcard, Harris family collection

What a collection they had—a prospector's pick, a snow shovel, a busted garden rake etc. No wonder that Mrs. Kerr was somewhat daunted by her futile efforts to undo Carpenter Creek's wild doings.

I sneaked quietly off to Bourne Bros. Store of which I knew that the great William Thomlinson was the very active manager. I introduced myself and Billy and I became friends almost instantly; I have never met a more friendly and likable man or one in whom a stranger would feel more confidence. Billy had had a most remarkable life commencing in a small farm (or croft) in the Cumberland dales ... At the time that I met him, he was the very life and soul of the camp; he was not at all "pushful" but so naturally a leader and so full of useful ideas and so straight forward and friendly that he was consulted in all public projects. He was most enthusiastic about the country and its climate and possibilities and the people. The boom was on and there was a great impression of prosperity. The mines were shipping quite heavily for such a young camp. The Payne mine was paying big monthly dividends, digging out valuable galena (sliver-lead ore) just as if it was coal. Just one immense block of galena found at the Slocan Star yielded a small fortune and it was lying right on the surface earth, "wash" as the miners call it, and then washing out the rich ore, just as they did gold elsewhere ...

Thomlinson immediately undertook to show me around and introduce me to friends, and he sold me some proper tools to start work on the big job ahead for Mrs. Kerr. So the next morning saw me pretty busy. I had borrowed a good crowbar from a neighbour and soon had a big pile of rocks rolled out of what began to look like a shaft for a well. Then I got a wheelbarrow and wheeled these away, and found some more hopeful soil ...

Kerr came back from Revelstoke and Billy came over to supper. Kerr was in great form; he had done well at court and had won two or three cases. He was in a brilliant mood. We talked for hours whilst Mrs. Kerr beamed her approval. Billy I found had a very keen appreciation of good literature. Burns and Cowper and Shakespeare, especially Cowper who was a Cumberland poet, therefore Billy's especial admiration for him. The Kerrs had a very noted and much disliked cat, they were cat mad, and could not bear that "Kitty Cat," as they called this pampered brute, should be thwarted in any way. The cat was a great trial to visitors, it would leap on the table and proceed to sample the contents of any dish it fancied. Eating after Kitty Cat did not appeal to most. Mrs. Kerr was a most awful housekeeper, she was full of brilliant ideas and most lofty sentiments and artistic yearnings, but she lived in one awful mess of rubbish and dirt. Dickens would have loved to portray her. She loved very bright colour and startling contrasts of colours (a riot of colours) rightly describes her sense of beauty. Kerr was neat and methodical in mind and nature ...

I spent three or four days in New Denver, where a farmer was a real curiosity, and most people regarded farming or gardening as almost an impossibility. I soon found, however, that there were others experimenting in tiny plots where there was far better soil than I had struck. The most advanced garden belonged to Mr. and Mrs. William Anderson who were most friendly and encouraging, for they had grown first rate vegetables and actually planted a few fruit trees. There was a whole lot to find out about the country and its climate. At that time, New Denver had only a few frame houses. There were many shacks and five hotels, all doing roaring business. The two largest hotels—the Newmarket ... and the Grand Central, a rough log building with a most pretentious front—ran busses [i.e. wagons, or stages,with seating and baggage space] up to the New Denver Siding to meet the daily trains ...

With almost everybody else in the whole country busy searching for mines, my peculiar and obstinate nature led me to seek for a small bit of good land, well situated with regard to the future population, that I could turn into a vegetable and fruit garden.

Where was to be the centre of the district and where would there be the most mines? My friends Thomlinson and Kerr and I talked these matters over, and found that the answer to the question was very difficult. Sandon, which was nine miles up Carpenter Creek, was the greatest centre of activity at that time. The Payne, the Slocan Star, the Wonderful, the Freddie Lee and several others, including the Reco and Surprise, were already shipping quite heavily. However, Silverton at the mouth of Four Mile Creek had shipped a big tonnage of fine ore from the Alpha and there was a wonderful showing of ore exposed on some claims just south, which had been named "The Galena Farm" because their proud owners claimed that you could actually plough out the rich ore lying on the surface. At Ten Mile Creek, very rich ore was being taken from the Enterprise and a wagon road was being constructed right up to the mine. At Slocan, Springer Creek entered the lake and the Slocanites boasted that they had the very richest ore that had been found in the whole camp ...

[Thomlinson] drew up a list of possibilities. Where the golf course now is. Allen and Cory's start of ranching down at Twelve Mile. The brick kiln below Slocan. The head of the lake and one or two other possible places that he could think of. He set down

Pack train, Slocan, ca. 1896–97. On first visiting Slocan City, Joe considered that it had no "appearance of permanence." New Denver Museum, Image 2001-022-006

particulars of them and offered me the free use of his rowing boat to get around on.

New Denver had certain solid advantages. It was the recognized government centre and with one or two mines of its own, even if rather small, it was bound to be the centre of the mines up Carpenter Creek and the distributing point for the whole district. But with the mountains rising abruptly from the lake and the forest covering the mountains, there seemed very little chance of finding land, even the quantity that I was after.

Thomlinson advised a trip to Slocan [City] and sent me to his friend George Henderson who had been one of the very first to enter the Slocan and who was now owner of the Arlington Hotel in Slocan. The hotel was new and rough, but comfortable—Slocan was booming with a mixed and wild population living in tents, cabins, shacks, etc.. [It] looked far too much like a gipsy camp. There were very very few decent houses or any appearance of permanancy. There were a few real miners who had come up from the Coeur d'Alene mining district.

Everybody spoke of the big open meadows a few miles down the Slocan River, but I soon discovered that these open meadows existed only because they were heavily flooded every spring and far into the summer, and the job of draining them or any part of them would be terrific, even if practicable. Also they were out of the way of any then known ore deposits, and ore deposits dominated the future of the district. I spent two days around Slocan City sizing up the situation. A small dairy had already been established on the west bank of the river. The enterprising dairyman had to bring his milk over by boat, and hay for his three cows had to be taken back by boat. The cows found bushes to browse on which helped out. Also there were some flooded meadows and mosquitoes galore. The absence of pasture throughout all other places was most noticeable. The forest trees had absolutely possessed the whole country except where the snowslides had swept them away and very rank vegetation was growing ...

After sizing up the opportunities offered by Slocan City I returned to New Denver. Talking with the miners and prospectors had convinced me that around New Denver, which had the backing of the mines of Sandon and Three Forks and also those up Four Mile Creek at Silverton, was the probable centre of the district.

However two of the real old timers, Mr. Johnny Cory, an old Englishman, and his partner "Long Allen", an American, had staked a ranch near the mouth of Twelve Mile Creek. Most of the creeks around the Slocan Lake had been named in accordance to the distance of the mouths from New Denver; not a very poetical method, but it had some advantages. They were prospectors from the great Coeur D'Alene mining camp in Idaho, but just at this time their interest lay in trying to open a townsite at Twelve Mile by making a new piece of wagon road to divert the traffic down Ten Mile Creek. They were busy building a hotel, for like all prospectors they were very handy men. They really had a very good location for the townsite and road, and the success of the Enterprise Mine, about nine miles up Ten Mile Creek, was directing a lot of attention to this part. E. W. C. Koch, the big contractor, had just built a road right up to the mine and his ore wagons were

The ss *William Hunter*, the first steamboat on Slocan Lake, n.d. Its machinery was packed in from Nakusp, and its body was built locally from whipsawn lumber. Image D-05938 courtesy of the Royal BC Museum and Archives

operating steadily, but he had a very poor dock and no proper facilities at the tiny wharf that he had built. Nor could it be extended at Ten Mile owing to the steep cliffs.

A great struggle was going on to decide where the future outlet for Ten Mile Creek would be. I was landed on the fine beach at Twelve Mile from the little steamer William Hunter, a wonderful little ship (whose engines were built in the Old Country and packed in, also the boiler, on mule back from Nakusp over 25 miles of rough trail) that had been built at the head of the Slocan Lake from lumber and timber whipsawed from the forest trees that stood there. It had been a wonderful piece of work and the boat was to prove a gold mine to its energetic builder and owner William Hunter ...

Thomlinson had sent word ahead to Johnny Cory of my coming, and Allen and he gave me a warm welcome, for the more they could attract to their landing, and get interested in their town site, the better chance there would be to win out. I was taken at once to see the ranch. It was at some distance from the wharf site and hotel, but it was a really nice piece of swamp land. They had drained it easily, with a ditch to divert the little creek, and even the first summer had succeeded in growing some very good vegetables (chiefly radishes). They had brought these up to New Denver, and they had sold like hotcakes. Of course there was only a very small garden patch that could be used so easily, but there was a considerable piece of swamp land with light brush on it that could be quite easily cleared.

It was certainly an attractive place and they were most eager to have me try it out. They would have allowed me to work out the small price for the land by working on their proposed road, which would be of the utmost importance in opening up the ranch. [Yet] it did not look so very hopeful as it was so far out of the way. Of course, if the mines up Ten Mile Creek flourished as was hoped and the road was put through it might not be so bad but it was a big risk.

Allen and Cory and I had lighted a fire and were busy cooking supper when we were hailed from the bush and a most important and original specimen of humanity came walking out. This was the celebrated Eli Carpenter, the French prospector who was one of the two first to find that the Slocan galena and zinc ores carried such unusual silver values. Eli was a very short man indeed and scarcely five feet tall, but enormously strong and active. He had been born in France at St. Omer, Pas de Calais (at the same town and time that his friend Charles Blondin, the famous French rope walker, was born). He was the playmate and assistant and rival of Blondin when a boy, and was the companion of Blondin when the latter gave his famous exhibitions of walking across Niagara Falls on a tightrope, wheeling a man across in a wheelbarrow.

Eli had given up his job with Blondin, or more likely his love of whiskey had broken up the partnership. Then he had taken to prospecting. He had tried placer mining etc. with various fortunes. He could exist most happily on food and under conditions that very few other men could face, and had a fine instinct for discovering mineral deposits. If only he had been able to keep away from whiskey, he could have been a very distinguished man. Eli had been prospecting up Twelve Mile Creek and had staked two or three claims on this trip. He had been out for over six weeks, living all by himself in a tiny shelter built in one day out of cedar bark, which he would strip off fallen cedar logs with the aid of his tiny prospector's axe, in about six foot lengths. He would start forth with a fifty lb. sack of flour and a little oatmeal, baking powder, tea, tobacco and salt and his rifle and trust to catching or shooting squirrels, fool hens, and deer, and perhaps above all porcupines.

Eli and I at once struck up a friendship. When he heard that I was prospecting for land he became very interested. He had heard of some very fine land and open meadows at the head of the Little Slocan River and he also wanted to go in to that country to prospect. Within a few hours of meeting, he and I agreed to join forces and take a trip together from Slocan City, where Eli had one of his queer little cabins, over the mountains and into the valley of the Little Slocan River. Eli was of course dead broke as per usual. I agreed to furnish supplies for the trip, which must take place at once as it was growing late and the fine summer weather might break up at any time. We sat on the beach at Twelve Mile, making plans and "chewing the fat" as talking in a friendly fashion was called. A tiny chipmunk came out from under some logs and we could see a huge lump sticking out of his side. Eli shot him with his old rifle and we examined that queer lump. When we pressed it an enormous maggot popped out, quite as big and very similar to the bots that we often find on the backs of cattle. It is the only time that I ever came across such a bot in a chipmunk.

Eli had left his boat hauled into the brush at Twelve Mile and next morning we left early, after saying good-bye to Allen and Cory. We pulled across to Slocan City where we bought our supplies, enough, Eli figured, for a week's trip. We included such superfluous luxuries as some bacon and oatmeal and sugar and rice. I expect that Eli felt that I was far too luxurious, but he certainly enjoyed them, and he certainly could cook, even with the scantiest outfit.

We rowed down the river for half a mile to where his cabin stood on the west bank. I sharpened up Eli's axe and cut wood etc. Eli was busy in some mysterious fashion, and before long announced that we were to have lots of fish for supper and breakfast. He produced some sticks of dynamite and caps and fuse. He broke the sticks of dynamite in half and very neatly greased the dynamite caps and inserted them into the dynamite, making sure that there was no chance for water to penetrate. He left about 9 inches of fuse and tied the charges up securely in a bit of old gunny sack with a rock to sink it. Then we went to a lovely pool that was close by. Eli lit the fuse and flung it carefully into the centre of the pool. A muffled explosion followed and a whole lot of fish came floating to the surface. We had the boat ready and soon had selected two or three fine trout and all the white fish (graylings) we needed. It certainly was a most wasteful way of fishing, but the white man is a most extravagant animal.

Eli made up our packs all except the blankets that we were to sleep in that night. Before daybreak we were up and we entered the brush and started climbing up the very rough mountain in the direction that Eli knew that we needed to go just at daybreak.[2] Eli certainly was a wonderful mountaineer. At times we would find a great log lying where it had fallen and Eli could walk logs big or small just like a goat. Sometimes we came to a small ravine and then Eli would look to see if a log was lying across which would act as a bridge and save climbing up and down. I funked many logs that he found easy. He took great care of me and was most companionable telling me of his wild adventures on the tight rope and the wild fights both with men and animals and storms and floods that he had been through. He still thought a great deal of Blondin though he admitted that they had parted after a tremendous row with whiskey and women taking part. We travelled fairly slowly, Eli carefully picking our way and of course breaking off chips of rock and examining each piece carefully in true prospector's style for who could tell where one might stumble onto riches in the Slocan country. We were a queer looking couple no doubt, Eli exceedingly short and almost hunchbacked, a grizzled little Frenchman with kindly twinkling eyes, and myself of the most lanky type of Anglo Saxon, fair and blue eyed and slim.

Eli carried his Winchester rifle. It was a short carbine and he fired either grouse shot or bullets as he encountered bears or grouse. All that day we only encountered one fool hen. Eli's rifle was in awful condition. He always meant to get it repaired, but whiskey came first when he happened to have money and the poor rifle was like a drunkard's wife—loved but sadly neglected. He aimed four times at that silly fool hen sitting quietly in a bush but three times his cartridge failed to explode. At the fourth attempt he shot that trustful bird.

We came to the open meadows as reported. Of course they were only open because they were flooded all the early summer. We began to look for a spot to camp but the difficulty was to find any decent water. There was plenty of dirty swamp water and plenty of good campsites. The beavers had been very busy and we saw lots of places where they had felled big trees and Eli thought of returning in the winter to trap them. The weather began to change and clouds to gather. Still we could find no good water. Then came a sharp shower of rain and the brush was very wet. Finally we had to make a hurried camp under some very thick trees. Cooking in the rain was not so good. However, Eli managed to turn out a fine supper and the tea tasted all right. It rained steadily and was still raining when we woke next morning. Eli had not found any mineral that looked at all interesting during the whole of the previous day's search. He felt that the whole rock formation was different from the country to the east of the Slocan River. I had enjoyed the trip immensely, but I felt that we were getting too far away from where there was much chance of any population and that land far back in the bush would be useless, so we decided to return for the wet weather appeared to have set in.

We started back following a different course but equally rough going. The weather cleared up and our clothes dried out on us before we got back to Eli's cabin, tired and hungry but satisfied. Eli was immensely impressed by the fact that I neither drank nor smoked. He thought that I could be trusted and he promised me when next he sold a mineral claim that he would come to me wherever I happened to be and deposit his money in my keeping. He certainly meant what he said but though he sold several claims he never turned up with the cash. Whiskey was far too strong for him. Carpenter Creek on which New Denver and Sandon stand was named after this most plucky little man ... [a] great and noble nature in spite of his weakness.

Back I went to New Denver a very perplexed man. Such a lot of possibilities and no certainty anywhere. There is a considerable piece of bench land overlooking the Slocan Lake between New Denver and Rosebery. It had already been staked as farming land and it was evidently rather light soil and the water supply was pretty small and uncertain. I went up to see it but do not think that I saw more than a small part of the land. An inexperienced man in a pathless forest can easily get confused and fail to size up the land properly. But after finding that the man who had located the pre-emption was not anxious to sell, he and I came out to the brow of the hill overlooking New Denver and I got a real look at what appeared to be a somewhat similar bench between New Denver and Silverton. I was determined to go and take a good look at it. So I set out for the place that was to become my home for almost fifty years. It was very easy to reach it for the public trail between New Denver and Silverton ran up onto it and along its outer edge. I got up onto a piece of flat land (comparatively so). Actually, from the brow of the hill overlooking the Slocan Lake there was a slight drop down to a queer swamp and at the far end of this swamp a beautiful little lake [Bosun Lake]. From this swamp a small creek ran, which however simply disappeared into the ground farther down.

[In another account, Joe wrote the following: "A young fellow named Smith had staked some land right between New Denver and Silverton. There was no road at that time between the towns but a fair trail had been cut out and the public telephone line had been slung from tree to tree along it. I walked up to see this pre-emption. About half way between the two towns I came on a small cabin, about 10 × 14, and there was a little clearing, at least the timber had been cut into four foot firewood. Two young fellows were at work cutting wood. They were Albert Victor Smith and his partner in the wood business, Bill Thompson. Good husky young men of whom Smith was easily boss. They had heard about me."]

I had already travelled along the trail from New Denver to Silverton several times. It ran through very thick forest and I had never imagined that there could be any extent of fairly flat land hidden in the dense brush. Now I deliberately made my way along the back of the flat just to the point where the steep mountain rises again to another little swamp and flat a few hundred feet further up. Soon I came on another small spring of water which also reached the flat and disappeared. Making my way farther south through an awful tangle of brush I came out to Smith and Thompson's little clearing and a rather bigger creek that made its way across the flat.

Here was land and here was water in what seemed plenty. The forest was tremendously thick and many of the trees quite large. It would take a whole lot of heavy work to clear it, but it seemed to have great possibilities. I knew Smith and Thompson by sight.

Smith had pre-empted the place chiefly with the object of finding a steady job cutting firewood. They had taken a contract to supply the *William Hunter* [a steamer] with 100 cords of wood delivered on the beach at $4 a cord. They meant to build a timber chute and slide the wood down to the beach in the wintertime. I returned to New Denver but came up again for a further examination of the place and I walked back and forth through the forest estimating as best I could what acreage there might be. A good deal of land was evidently very light and there were many big granite boulders scattered around over it. Some small parts appeared to be good black soil with excellent chances to get water on them. The position was magnificent right between New Denver and Silverton. Already there was an agitation for a public road to connect the two towns and roads were certain to be built throughout the mining camp before long.

I approached Smith and Thompson and we were not long in making a deal. I had to pay them to vacate the pre-emption and then I had to restake the piece of land. My original stakes were not placed to the best advantage and afterwards it was necessary for me to locate a further fraction of a claim so as to take in the land along the Lake.

I became the owner of 245 acres of very mountainous land, less than 20 acres of which was really fit for cultivation.

[This in another account: "A. V. Smith showed me all over his preemption. The position was certainly almost ideal half way between the two little towns. And here were no less than three little creeks running across the long narrow flat. There was also another flat above on which some very fine cedar, white pine and fir was growing in and around a regular cedar swamp.

"There was a small piece of good land close to the cabin which looked as if it could be drained and would make a first rate garden. The bench land on the lower flat was very light, regular bench land which would need a lot of work and manure to make much of it.

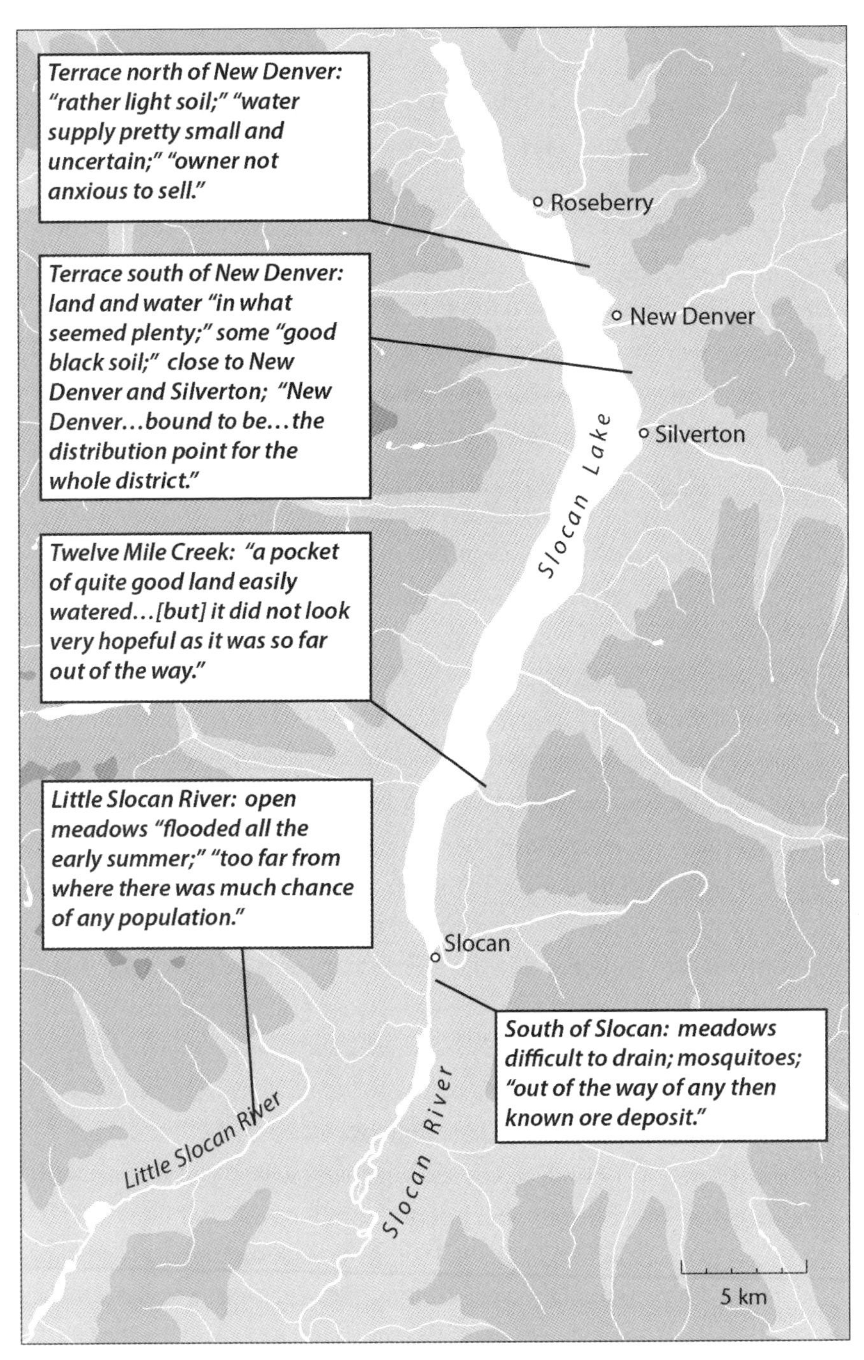

Looking for land in the Slocan, 1896. The locations that Joe considered for his farm are indicated, and his evaluations of them summarized. Map by Eric Leinberger

It was covered with a tremendous growth of timber. The timber made it very hard to get a real estimate of the amount of land that was really fit for farming but I soon satisfied myself that there was sufficient for my purposes; though even what was included in the flat was really only somewhat flatter. It was all hilly. There were a lot of rocks and a few quite big boulders scattered about, but in my enthusiasm, such trifles did not greatly worry me. Smith said that he was prepared to sell out and after some bargaining I gave him his price of $1000.00 and he was to finish his contract of delivering 100 cords of wood down on the beach for the steamer *William Hunter*. He and Bill Thompson were to stay with me in the cabin until their contract was finished."]

My buying the land caused some little stir and a good deal of land was staked in the neighbourhood around the little towns. Another serious difficulty arose as Jim Delayney, the smart proprietor of the Grand Central Hotel, had located two mineral claims, fifty acres in each, just before I had bought out Smith. They were utter wildcats: no mineral in sight on either and no real ledge exposed. This was against the new British Columbia Mineral Act which requires that a true mineralized ledge must be exposed on every claim located. But as I should have to prove that the men who located and recorded the claims had not found anything that they were justified in thinking was a ledge and mineralized, and such a negative could hardly be proved, the law was of no use to me so on Kerr's advice I had to buy the two claims for $700 cash.

I have not mentioned that there was a pretty fair log cabin on my land. It was about 12 × 16 ft. and had been erected by a man named Sammy Sturch.

Smith and Thompson cut their firewood just around the front of the cabin. They built a wood chute, from very near to the place where the iron gate in front of the ranch house is, and hauled the wood there and slid it right down to the lake shore. It worked very well but certainly they earned their money for it.

Both were expert woodsmen and hard working fellows and I got on well with them and learnt a whole lot about the different sorts of trees and their values from them. The timber on the ranch at that point was very thick indeed. They cut their 100 cords of wood on a very small part. They made a good job of it, leaving the brush in windrows ready for burning and roads between brush piles for working a horse through them. I set to work at once to enlarge the opening in the trees. The cabin was situated right by the trail which led from New Denver to Silverton. The country was full of people all very much on the move. As many as 40 would pass the cabin in a day. The trail was about four miles long and the cabin about half way between towns. There was a fine little creek of beautiful water crossing the flat only a few yards away. A good-sized pan sitting on a convenient stump was our rather primitive lavatory. We took our Sunday tub sitting in the family washtub right in front of the cabin. We just upset the tub instead of pulling the plug as city people were accused of doing.

I started work at once, so as to make the most of the short time to winter. Smith had taken just the big trees and I chopped down all that was left and I managed to get a good deal of the timber burnt right off by November. Probably nearly an acre, with stumps as thick as the hair on a dog's back and lots of rocks right on the surface and lots more that the plough was to show up. I also dug the ditch to drain the swamp.

Life was very simple and happy. After Smith and Thompson had rashly left me to do the cooking on one or two occasions, they wisely decided to do all the cooking themselves and I chopped wood for the stove and did most of the washing up.

The place had been located originally by a man named Sammy Sturch who had left the country. Sturch was one of the locators of the Galena Farm Mine and he had got a considerable sum for his share. He was a house painter by trade and he had built the little cabin. One of his paint pots had been left just outside, with a little bright blue paint in it and a small brush. One of Sammy's friends had come there to see him and found him absent, so he had taken Sam's brush and had written in big, bold letters on the door. "Sammy Sturch, why in Hell d'ont you stop at home and mind the ranche??" This writing remained there for some years, until the cabin got burnt down. Smith had given Sammy a few dollars to relinquish his pre-emption and now I gave Smith $1000.00 to relinquish his claim in my favour.

As the weather got colder we began to find the cabin most awfully cold. It had only a shake roof, with innumerable ventilators. However Smith got plenty of newspapers and lots of paste, we papered even over the rough shakes, and the difference was tremendous.

Then I went back to Vancouver Is. to take some well-earned rest and to visit old friends and pack up for my return very early in the spring. I travelled by Nelson, then to Robson and up the Arrow Lakes. A drunken passenger reached into the window of the purser's cabin and stole his great coat, he walked aft, and then feeling that he was spotted, he took off the coat and threw it overboard. There was a great row and the Captain decided to put the man off at Deer Park. We arrived there, and there was just a wharf, not a sign of a house in sight. The man was put ashore but some of the passengers objected to putting off a man in a dark, cold night where he might even perish. They said that the Captain would be responsible and finally he saw the point and the man was allowed back on board. We ran aground in the Narrows, but after sticking for nearly a day, we managed to get clear.

I had a grand time back at Duncan where of course very many thought that I was mad to go up to the wilds of West Kootenay. I sold my old mare Topsy and I bought a very fine team of rather small Clydesdales—Rock and Dick—two noble horses who were to do a tremendous lot of hard work up at New Denver. I got a stumping outfit, which was very simple and effective. It consisted of three lengths of steel wire cable, with very strong hooks at each end and three good snatch blocks, and a short piece of heavy chain for hooking the blocks to the stump to be pulled. I made the mistake of getting the whole outfit rather too heavy, it was stronger than necessary; however it did good work for many years. With the single block, we doubled the horses' pull, by hitching the second block to the fall of the first, we again doubled the purchase, and the third redoubled that pull, making the total purchase equal to 16 horses. The ordinary stump pullers were even more powerful, as they were mostly made like winches, but on uneven ground they were very awkward and dangerous to horses and men. I got a new plough and harrows and my old friend and neighbour Micky MacFarlane made me a set of bob sleighs, in payment of an old debt. These were not a success. Altogether with Cleverley's help we made up a carload.

Of course I brought up my young friend and helper Arthur Cleverley. He was a boy from my own little town of Calne in Wiltshire ... His father came to my dad to ask him to allow his son to come out to work for me and I had been consulted and agreed ... I had my old friend the Bosun, alias Charlie King, a runaway sailor from the Royal Navy, living with me when Cleverley arrived. The Bosun was one of the many lovable old drunks that I have known—awful when drunk but exceedingly good company when sober and a good workman ... We all knew when Bosun was getting over his drunk. First he would have a cup of tea and then he would fill his pipe, and if he could smoke it he was pretty well on the mend. Then he would ... grow as happy as a clam at high tide (his favourite simile) until another bout of drinking was due ... He was a witty old boy and always clean in mind and body. He had been apprenticed to a baker whilst a boy, before he joined the Navy, and was still a great hand at making hot biscuits, buns, and sinkers, as he called his delicious dumplings. Bosun and Cleverley and I became great friends. Cleverley came from a family of celebrated cooks and when Golly [the young Chinese cook] left he took over the cook's job. He knew a great deal about gardening and soon became a very handy man ... [H]is and the Bosun's stories of queer doings, and "high life below the stairs" would have filled volumes. [See chapters 1 and 3 for more on Cleverley and the Bosun.]

We had some final festivities to bid farewell to my friends in Cowichan and then Cleverley and I set forth. Old Bosun was very sorry to see us depart as my home in Cowichan had been a great place of refuge for the old boy, however I promised him that if we got settled in the new home I would send for him and I did bring him up after a time and he was quite happy up in the Slocan and a very considerable help.

It was an awkward journey, with the crossing to Vancouver and the changing of the cars etc. Then the long ride up in the boxcar, sleeping on the baled hay, however we were young and the Prophet [Cleverley] was good company and saw the funny side of everything. We arrived in Revelstoke and I got a little more hay and some oats and then on to New Denver Siding and arrived to find about two foot of snow on the ground. I had given Smith a contract to build a log cabin for a stable and Sam Watson and Smith another contract to build a sleigh road, up from the Silverton–New Denver road. So directly the car was spotted up at the Siding [New Denver Siding], we set about moving our stuff. We got the horses off at Clever's [the local butcher's] cattle chute and pretty soon we had the sleigh out and the team hitched up and we put on a load and set out for town in great style, though I felt very ashamed of the rough home made sleigh.

However Rock and Dick in fine new harness were a fine sight. I had never driven a sleigh before (except little cutters). I knew absolutely nothing about deep snow. I got down to town and wanted to cross from one street to another and never thought of hidden stumps and started across over what looked like a fine level field. In a moment the horses were floundering in the deep snow and the sleigh was hitched up on one of two stumps, which were hidden by fresh snow on top of the old crust. We managed to get clear after a big struggle and I determined to keep clear of unbeaten tracks in future. Pete Angrinon [a tall, powerful French Canadian whom Joe much admired, owner of a livery and freighting business] saw how very green I was, and kindly advised me not to take the horses into deep snow, as they would overact and pull their front shoes off and lame themselves.

Then we set out for the ranch. Of course my new road was quite unbroken. Following Pete's advice we did our best to break the trail by walking ahead and trampling the snow, but it was very heavy, slow work. However we made it at last and got to the new barn. It certainly looked a pretty rough, cold place: an unchinked log building with a shake roof. Cleverley went into the little cabin and set to work to make supper and I looked after the horses. Sure enough their front shoes were almost wrenched off and they would need reshoeing first thing. I stuffed up the worst chinks with hay and did what I could to make them comfortable. We had no light as our lantern had been left behind at the car. However we made out as best we could, and had lots of fun amidst the discomfort. And we managed to put in a fairly good night.

Next morning we were up and out as soon as possible and I took the team into town to get them reshod and to bring back another load from the boxcar. I went to Pete's barn and hunted up a most grumpy old blacksmith who had been on a spree the night before and after much delay he fixed up the shoes, and I managed to bring back a fair load. After that the hauling went pretty well, and we got all our belongings safely to the ranch, and before long we had the barn chinked up with hay etc., and life was easier. I got into trouble once or twice more with the team in deep snow that spring. The conditions were so new to me and the way that we came on deep snow, even when it was gone elsewhere, was astonishing. It was an unusually late spring, after exceptional snow. I feel I was lucky that I did not ruin the horses under those conditions. We began to work then at hauling logs and moving the last of Smith's wood to the head of the chute and I also got the free gift of John Taylor's (the New Denver milkman's) manure pile and whilst sleighing lasted I hauled a good many loads. Cleverley and I established a small hot bed, digging out about a foot of snow to do so. It gave good results. Cleverley was certainly a very good gardener. However there was very little work for the team.

How we worked all that spring. We were dreadfully crowded in the little cabin. I had a young fellow from Chemainus named Allen Porter, brother to one of the numerous flames of the very inflammable Cleverley and a very capable workman, to help, and also Old Bosun so we were quite a family. Clearing the land proved awfully hard as the stumps were so green and tough. It was more than double the work of clearing land when no time is allowed for the stumps to rot a bit. However, we managed to get about one third of an acre ploughed and really clear of rocks and stumps. Then the manager of the Galena Farm asked me to hire them [out], the Team, for the purpose of hauling up supplies etc. to the mine. It was a good job and I was glad to let them go for all that summer. They were well looked after, it was steady work and not too heavy, and it left me free to go ahead cutting down trees.

With the manure that I had hauled on the sleigh we were able to get our one-third acre into pretty good condition for new land. Cleverley ... did most of the gardening and we had a pretty fair garden. We had brought up some rhubarb plants, and a few sticks of rhubarb was our first produce sold off the ranch to Mrs. Estabrook, wife of the Captain of the *William Hunter*. She was also the Postmistress and a splendid woman, most kind to all the young fellows. Cleverley also grew a fine patch of radishes just outside the cabin, and we sold them in Silverton for the 24th of May celebration.

There was no water for irrigation. We had cleared the piece of land just above where the ranch house now is [Granny's Garden].

The little valley with the ditch containing the creek lay below it. Water was necessary and I made a trip to the Head of the Lake in Billy's boat to get some lumber to build a trough across this little valley. I got 1 × 6 boards from Hill Bros. We made a little "v" trough of these boards and carried the water over to the highest point and from there irrigated the garden. It worked well. The first night after it was up a porcupine appeared and drank out of our trough—the very first that I had seen but I have seen plenty since. Iron pipes would have been far better, but it was harder to get them then.

As we began to produce a few vegetables I bought a boat and started to hawk them around. There was not much difficulty in selling fine fresh vegetables, but the labour of packing them down to the boat and of rowing to town every Saturday and carrying them around town was very heavy. I should have done better with a little packhorse and pack saddle; however there was scarcely any pasturage at all and a horse would have been a considerable problem. Towards the end of the summer I had regular days at New Denver and Silverton and there was a good deal to sell. Sometimes I came home dead tired but happy, with over twenty dollars in my jeans. Sometimes people would walk up and buy something and carry it away.

We had a tremendous crop of vegetable marrows [zucchini-like squashes], grown between the stumps. We cut over forty fine marrows on several occasions, and how sick I got of trying to sell marrows on the glutted market. I had enjoyed selling our produce and some of my customers were most kindly and it was a real pleasure to deal with them. Others were mean and suspicious and would haggle and try to beat down prices. What a lot of human nature showed up.

We were a very merry crowd at the ranch, very carefree and very hopeful although things were not going extra well in the Slocan. Silver was down and a lot of people began to move out.

Allen Porter left and I got two young fellows to come and cut firewood for sale. They boarded with me. One an Englishman named Armistead (son of an Anglican parson) and the other a Canadian from near Guleph named Brierley. They cut a lot of wood for which I found a ready sale next winter.

Some of us slept in the barn on bales of hay, making a hay bed on top of the bales. The road between New Denver and Silverton was impassable for wagons and barely good enough for a sleigh road.[3] When I got the team back that fall I had to drag the feed back from town on a go devil [a simple one-horse sled used for hauling]. We determined to build a big log cabin as the weather got colder, to house the gang, and we had a great time and finally made a pretty comfortable and large log cabin, with rooms upstairs.[4] It was a great day when we had it ready and we were able to move in. We lined up the inside with shiplap and after a time, the walls were papered, making it very comfortable for bachelors.

A destitute and helpless young Scotchman who was hanging around New Denver now joined the crowd. He came just for a night, but had nowhere to go and so stayed on always looking for a job, but hoping most sincerely not to find one. He had started to be a doctor but was too lazy ever to qualify. He had plenty of brains and could talk like a streak. Kerr said "I consider Forbes fully capable of taking charge of any continent except the one on which he happens to reside." Of course he did a little and he was very useful in going to town and fetching the mail and small parcels of supplies and he was very amusing. He it was who gave Cleverley his nickname of the Prophet, which suited him to a T.

Puck was a very important part of the family. He had the most extraordinary grit. He would hang on to a bit of rope and you could swing him round and round. He would dive in the water for rocks thrown in and he would go for any dog, big or small, that came along. I think that he was the most popular, best-known dog in the country. He was forever putting wild cats up into trees in the wintertime and he was always getting himself porcupined or skunked. Finally he dragged himself home, mortally wounded by a revolver bullet we thought. Great was our grief.

I had a very busy and successful winter, hauling firewood mostly. I sold my own and I hauled a lot for other people. The very crazy and drunken young fellow who was running the electric light plant (called Bull Dog McDonald) managed to do me out of quite a bit of money for hauling, but he also managed to defraud most of the citizens, so at least I was in the fashion. Otherwise I got paid for practically all I did. I used to haul in a load of wood and would haul out a load of manure, so that by spring I had a fine pile ready for the very limited amount of land that I was able to clear. I also filled several ice houses.

We tackled the piece of good land lying in the little creek bed next. It was tremendously hard going, as the timber had been exceptionally heavy and thick. I should have been better off not to clear so thoroughly, but to have cleared off the brush and burnt everything clean and then sown grass & clover for pasture, to enrich the land for hay.

Of course we had more produce to sell next summer, but the market was not quite so good.[5] People were harder up and they had many more gardens in town, however things were pretty good.

JOE'S FINAL REMINISCENCES ABOUT CLEARING AND FARMING ARE SKIMPY. Perhaps he was running out of steam, perhaps he had said what he wanted to say about the beginning of the Bosun Ranch. Beginnings are usually more intriguing than their aftermaths, and the principal beginning on the Bosun Ranch in 1898 was the discovery of the Bosun Mine. That story is in chapter 4.

3 THE EXPANSIVE YEARS: 1898–1918

JOSEPH (JOE) HARRIS'S REMINISCENCES, WRITTEN BETWEEN 1943 AND 1951, end at 1898, and there is nothing like them covering the next two decades, when the ranch fields were cleared, an orchard planted, a log building subsumed in a sprawling house, and four children raised. Nor are there family letters from these years. There is a speech in 1909 by Joe about common errors when establishing orchards. Otherwise, I have only family stories, a short handwritten account by my father about the early ranch, my notes from several conversations with him in 1972, my cousin Nancy's writings about our grandfather, and notes from an evening's conversation with Uncle Sandy in 1985. That is all. They permit an outline of these years, not more.

IN EARLY DECEMBER 1898, JOE STORED A SMALL POTATO CROP IN THE LOG house he and his crew had built the year before, and returned to England to see his family, and, one suspects, find a wife.[1] On the Atlantic liner, he met and fell in love with a dark-haired Scottish beauty from Fyvie, a farming village northwest of Aberdeen. At Liverpool they parted, he for Calne, she for Aberdeen. Completely smitten, he wrote to her several times a day until his mother, a woman of some command, wrote to a young woman she had never seen to ask her to come to Calne, marry her son, and put him out of his misery. She did. In the early spring of 1899, J. C. Harris returned to the Slocan with a wife.

In regards to Margaret Raeper, the wife in question, only an uncertain outline of her early life survives. Her father was a saddler and harness maker, and Margaret one of twelve children. She left home as a young woman, took a course in dressmaking in Aberdeen and then seems to have been employed either as a governess or a dressmaker. As a governess, she spent almost a year in Malta with a British military family, returned to similar employments in England and Scotland, then found a position with another military family in northwestern India. After several years in India, the climate undermined her health, and she was sent home, where she recovered and opened a small dressmaking and seamstress shop in Edinburgh. One of her customers was Mrs. Harper, the young, attractive, and clothes-loving wife of John Harper, president of the American publishing firm Harper & Brothers. Finding Margaret charming, Mrs. Harper asked her to become her personal companion; apparently they became close friends. From this perch on a glittering world, she travelled with the Harpers in France and Italy, meeting many celebrated people. At some point, Mrs. Harper brought Margaret to New York and set her up in an exclusive dressmaking shop, from which she was returning to visit her family in Fyvie when she met my grandfather. She was thirty years old.

Margaret Raeper, ca. 1898. This is the only surviving photo of my grandmother near the time of her marriage. She was thirty years old, accustomed to some comfort, and could not have imagined the life that awaited her in the Slocan. Harris family photo

In Calne, to which Margaret was summoned, she encountered a small town industrial English version of the affluence she had known with the Harpers. The Harrises were a prosperous industrial family, by this time owners of one of the largest meat-packing factories in England. South Place, the family seat, was a large house well staffed with servants. And my grandfather, handsome, fashionably dressed, and well-mannered,

Joseph Colebrook Harris, ca. 1898. A very English picture of a prosperous young Englishman, one that ill describes his relocated life in the Slocan. Harris family photo

obviously an offspring of privilege, told her that he had an estate in British Columbia. She may have been looking for another adventure in an already adventurous life or, perhaps, by marrying one of the Calne Harrises, for comfort and financial security. She could never have imagined what she would find. Years later, she told my mother that as she and her husband got closer and closer to the Slocan, the estate got smaller and smaller. When they finally reached it at the end of a jarring wagon ride from the wharf at New Denver on an improbable, end-of-winter road, it became a log cabin stinking of rotten potatoes in a tiny mountainside clearing. Had there been anywhere to go, she said, she would have gone.

There wasn't and she stayed, although it seems to have been a close call. In retrospect, even my grandfather acknowledged the problem. "It is not fair or right," he wrote a decade later, "to dump a woman down in the bush without near friends or neighbours, as the farmer's wife often is."[2] Years later he admitted that "the big log cabin had been a pretty good home for a lot of bachelors but was quite a shock to a girl from New York and Paris. However my wife had been much cheered by the name Slocan City and had determined to do her shopping at that big centre [then even smaller than New Denver]. It looked pretty hopeless ..." My grandfather hired a carpenter to enlarge and civilize the log cabin. Perhaps that induced her to stay, or perhaps affection for my grandfather. Or perhaps because she was pregnant.

Late in 1899 during a difficult birth the doctor told my grandfather to choose between a baby and a wife. The baby, a large and apparently healthy boy, was buried at the top of the wagon road leading down to the lake. My grandmother, a short, slight woman married to a large man, starved herself during her next pregnancy; my father—Richard "Dick" Colebrook—weighed only four pounds at birth in December 1900. The other children—Alexander "Sandy" Leslie, Heather Elizabeth, and Thomas "Tod" Oliver Douglas (also known as Joe)—each weighed more.

At the ranch, Margaret became a farm wife and mother. Although for many years there was a succession of Chinese cooks, each much loved by the Harris children, and help from the Old Country—Kate, a young Aberdeen girl who came about 1904 and stayed until she married a couple of years later; Dora Clemson, who was interviewed by the family in Calne, arrived in 1909, stayed for years, and became a fast family friend; and Ella Harris, an English second cousin of Joe's, who came about 1911—my grandmother herself seems to have taken on an endless round of work. "I thought," she once told my mother, "I was going to have an easy time when I married a Harris, and I never had to work so hard in my life." Her talk about the past, Mother wrote, "is always work, work, work."[3] There were British standards to maintain: starched white collars, for example, worn by children walking two miles through forest and snow to school. When wringer washing machines arrived, my grandmother, who had spent much of her life with washboards and flat irons, said they just made her mad. She also looked after the flower garden, the greenhouse attached to the front of the enlarged farmhouse, and many farm chores. By all accounts she was a cheerful, chuckling, story-telling person, and probably a good deal more practical than her husband. My uncle Sandy often said that his father got the credit but his mother held the place together. Significantly perhaps, her youngest son, Tod, working at the Nickel Plate Mine in Hedley in the 1930s, identified himself as Scottish. I just barely remember her in the early 1940s, a small, bent woman shaking with Parkinson's disease. We know that she came to love the home she created at the ranch and that she was the centre of it, but not much more. Some of her children's rhymes still circulate in the family:

> Twa little doggies to the mill they ran,
> This way, that way, this way, that way.
> A lick out of this wifie's pioch,
> A lick out of that wifie's pioch,
> A loop o'er the glade
> And a drink from the dam,
> A puckle for the miller
> And another for his man.
> Then hame again, hame again,
> Loopy for span.

There was money and momentum on the Bosun Ranch during these years. Joe's father died in 1909, his mother a decade later, and from each, apparently, he inherited some nine thousand pounds. A good fraction of these sums went into the ranch. A steady stream of relatives and others arrived from the Old Country, some of them in response to advertisements placed in English papers inviting young men to learn the art of farming on the Bosun Ranch. For a time relatives sent care packages of English clothes: knee britches for Joe, English schoolboy suits (much disliked by their recipients) for his sons. The ranch house was enlarged and landscaped, farm buildings erected, and a mixed commercial farm, with an orcharding emphasis, created.

By the time of my father's early memories (1904–5), the main field of the ranch was cleared, although a stumping machine was still in use. Dynamite, generously employed, also served. As a small boy, my father watched from a ranch house window as a large fragment of a stump, blasted out of the ground by my grandfather (he was not good with dynamite), sailed through the air and landed on the roof. The field below where my parents' log cabin would be built (see chapter 7) was cleared about 1910, and the Far Field by Billy Freed, who lived there in a house belonging to the Bosun Mine. The steep bank above the road between New Denver and Silverton and below the ranch house was cleared about the same time by two Swedes, Erickson and Johnson, Canadian rock-drilling champions,

The Bosun Ranch orchard, 1907, a photo to send to the family in Calne. It shows a thriving orchard and hides its underlying dilemma, the lack of a market for fruit. Harris family collection, photo by R. H. Trueman

who also lived for a time on the ranch. They chopped wood to keep fit and felled trees and cut them into cordwood with axes. They also practised drilling on one of the ranch's granite erratics while my father, then a small boy, counted down the last seconds of each minute, after which striker and drill holder switched, without, if possible, losing a stroke.

First and foremost, the ranch was to be an orchard, and the entire main field was planted in fruit trees. There were some thousand trees, most of them apples (Wealthys, Gravensteins, Yellow Transparents, Duchess of Oldenburgs, Lord Suffields, Baldwins, Russets, Kings, Wagners, Cox's Orange Pippins). There were also plums (Peach plums, Italian prunes, Damsons, Lowlands, Yellow Eggs), and pears (Bartletts, Clapp's Favorites, Flemish Beauties), and more than an acre of cherries (Lamberts, Royal Anns, English Morellos, Olivettes). Initially at least, the market was assumed to be with the mines, and different species and varieties, ripening at different times, were intended to supply them through a long season. Years later, Sandy said that his father "made a beautiful job of what he started to do."

By 1909, Joe himself was not so sure. The assumption around which he had planned the ranch—that the local market would absorb its produce—had proved wrong, and the advice he gave that year in a talk to the Farmers' Institute in Nakusp, the only surviving record of his thoughts about his emerging farm, is basically a critique of his own efforts. He thought he had tried to do too much too quickly. He had been too isolated and too busy clearing land to make use of the experience of the few neighbours he had. He thought he had cleared too much land: "we want small, well-tilled, highly cultivated farms, not big, half cultivated wildernesses." He thought he had planted too many varieties of apples: a grower should plant only one or two varieties, a district only two or three. In this way, the scale of production "will command respect and attention from the express and railway companies," and dealers would come to know where they could be supplied. His own impatience to clear land and establish a farm had created far too many mistakes:

> [WE HAVE BEEN] in too great a hurry to plant our trees and get things started to grow. The loss of a year seems a terrible thing; therefore we order more trees than we are really ready for. It does not pay. The trees may live and even turn out well, but you may be sure they would have done better if the ground had been thoroughly worked beforehand. Don't plant amongst stumps if you can possibly avoid it, and don't plant on steep hillsides; you can grow fine fruit both amongst stumps and on very steep banks, but it is going to take a whole lot of work to do it, as I have proved to my sorrow. Don't plant near buildings, except cherry trees to use as shade trees. Plant a good distance apart, it is hard to work with horses in an orchard where the trees are under 20 feet apart; 30 feet apart is better; trees need light and ventilation to flourish. Get clover growing on your land as quickly as possible; it will do more good for your land to get it mellow and productive than anything else. Build good fences before planting trees.

These, he wrote, were "only specimens of the mistakes I have made, and by no means exhaust the list." Some could be corrected fairly easily, but others, particularly those bearing on the variety and spacing of fruit trees, could not.

At first, according to my father, there were no orchard pests, but after a few years coddling moths and others arrived, and with them the need for three sprayings a year, usually lime sulphur or Bordeaux mixture. In my father's memory, the ground under the fruit trees was cultivated. Irrigation water, necessary in a summer-dry climate, came from a small dam (it remains visible) at the head of the narrow draw above the barn. Pipes were dug in, stand pipes erected, and water distributed from them by hoses and sprinklers. A small sawmill built just across the creek from the barn was powered from the same source. Boxes were assembled and fruit packed in a packing house in the middle of the orchard. Each box of ranch apples was labelled "Kootenay Apples, produced by J. C. Harris, New Denver, BC." Sandy remembered picking apples at 4:00 a.m. for morning pickups at the lake by the SS *Slocan*.

There were always horses, cows, pigs, and chickens, and often sheep and ducks. A large vegetable garden filled the low, mucky, south end of the main field. Behind the ranch fields was forest with merchantable timber. So composed, the ranch produced many commercial products: all the tree fruits identified above, vegetables (particularly potatoes, cabbage, carrots, turnips, peas and beans), eggs, chickens, butter, honey, ice (from the Bosun Lake and Loch Colin), cordwood, mine timber, and saw logs. Year after year, it won the prize for the best general display, which might include trout from Loch Colin and socks knitted on the ranch from wool from ranch sheep, at the New Denver Agricultural Fair. According to my father, there was delight in as much variety as possible.

These various productions required endless manual work. The ranch workday began shortly after 5:00 a.m. and stopped twelve hours later, with breaks along the way for breakfast, lunch, and morning and afternoon tea. Sunday was a day of rest. Joe worked as hard as anyone while directing a fluctuating workforce of relatives and hired hands from the Old Country, and local men down on their luck who often found some work at the ranch. Among this diverse workforce several names stand out: Arthur Cleverley, the Wiltshire lad who accompanied Joe to the Slocan in 1897, and remained on the ranch for many years. Charlie Raeper, Margaret's brother, who came about 1901 and built the three-room cottage that later became Sandy's house. Ernest Hawes, a small, precise Englishman, who looked after the chickens. Colin Harris, a Wiltshire cousin, who came out in 1908 and made the pond named after him (Loch Colin) with horse and dragline. Heinrich, a German who looked after the ducks. Two English public school men, King Brammel (Mill Hill) and Tom Hepburn, who were on the ranch for a few years before the war and contrived to play cricket in New Denver. Lionel Henley, a Harris relative who worked on the ranch before the war, then enlisted and was killed. Tommy Rankin, a graduate of Edinburgh University who worked on the ranch for two or three years, became a schoolteacher in New Denver, and was also killed in the war. The Swedes, Erickson and Johnson, who lived for a time in a house in the Far Field and chopped wood to tone their muscles. The Chubbs, a hard-up family with seven children from down the Slocan Valley, for whom Joe enlarged the cottage Charlie Raeper had built and employed the father. At some point, Joe hired a ranch foreman, Fred Brady, a graduate of Ontario Agricultural College, to manage this labour, an arrangement that did not work and was quickly terminated.

Most of the farm buildings in 1914 are shown in the photo opposite. Adjoining the ranch house at its northward end was a large, steep-roofed building that served, apparently, as a milk and ice house (for separating and refrigerating milk) and, upstairs, as accommodation for hired hands. At the right of the picture, and at the brow of a low hill, was a barn and beside it another building, perhaps a stable or chicken house. There was a chicken house somewhere on this hill—my father remembered the night a weasel slaughtered fifty chickens. Beyond this picture and against the mountainside, a second barn was built between 1909 and 1914. Across the creek from the new barn was the sawmill. Across the creek from the ranch house were several small buildings, one of them a pigsty.

The largest and most impressive building was the ranch house itself. The two-storey log structure built in 1897 had become encased and almost

Farm buildings near the farmhouse, Bosun Ranch, 1914. Although the early Bosun Ranch was primarily an orchard, there was also a large vegetable garden and all the common farm livestock. Most of the farm buildings that supported this varied production are shown in this picture. The principal barn, used for dairy cows and hay and located at the boundary between field and mountainside, is outside the picture. Harris family photo

invisible within a sprawling, Carpenter Gothic house with, by some counts, eighteen rooms, the largest house in the Slocan Valley. It seems to have been built over several years, beginning in 1899, principally by Andy Wallace, a carpenter from eastern Canada, and to have reached its final form in the winter of 1909 when, to facilitate enlargement and reconstruction, the family moved to New Denver. Certainly, English money paid for it. A greenhouse intended to start plants and provide ferns and other greenery year-round was attached to the front and heated by radiators supplied by the house's hot water system. Sandy said that it was a lot of work for little return, but that it grew some "smart things" (lettuce for example), and that his mother enjoyed showing the only greenhouse in the valley to visitors. A guest room and children's play room were in a rear extension. A porch or verandah ran around two-thirds of the building. A basement, under about a third of the house, was accessible by an external door and a wood chute. From the front, this was an attractive and, for the Slocan Valley, an imposing house, but, like others of its day, it was uninsulated and, because of its size, particularly hard to heat—it had fourteen external doors, at least one of which, Uncle Sandy often said, was usually open, even in winter. Its furnace and hot water heating system consumed a prodigious quantity of firewood—over forty cords a year, much of it cut by hand. For several years, a small power house and Pelton wheel on the creek just above the Silverton–New Denver road furnished a brown electric light turned on by a valve on the back verandah.

The landscaping around the house, which survives, battered by time, to the present, dates from the same years (1907–9) as the enlarged house. It was laid out by Herb Thomlinson, described by Sandy as "the best landscape gardener in New Denver." A rock retaining wall, built at the top of the steep slope to the lake, created a flat, elevated space, several feet higher than the flower garden my grandmother had originally tended there. This space was terraced, and most of it made into lawn using wild grass brought from the shores of Bosun Lake. Heinrich Christianson, who had served two years in the German army, did the cement work: the concrete gateposts that mark the upper end of the ranch road (coins and other items of the day are embedded in one of them); the concrete steps, path, and gateposts leading toward the orchard; and the concrete border between lawn and orchard together with the wire fence embedded in it.

Running through this space in a broad, gravelled sweep, and then narrowing through the gateposts, was the upper end of the ranch road. Beside the right gatepost (heading out) was a chestnut tree, on the terrace a mountain ash, and below the retaining wall and near the house a Cox's Orange Pippin. Just beyond the wire fence was Granny's relocated flower garden. A measure of elegance had been created where ten years before was only the charred aftermath of clearing. Possibly, the concrete steps

The ranch house on Bosun Ranch from the front, n.d. but probably about 1910. With some eighteen rooms and eleven exterior doors, the ranch house was probably the largest private house in the Slocan. It may have been built partly to please my grandmother but, like everything else on the ranch, it entailed an enormous amount of work, much of which fell to her. Harris family photo

leading to the orchard were simplified overseas reproductions of the steps between different terrace levels in Joe's boyhood garden at South Place (see photo, page 64).

A mahogany cabinet brought over from South Place stood in the parlour together with a Heintzman piano, a Victrola phonograph, a Morris chair with an adjustable back, a cabinet for sheet music, and a green Wilton carpet with a floral design. The fireplace, faced with green tiles, had fleur-de-lys firedogs made by the blacksmith at the Bosun Mine. Above the mantel, which was supported by two wooden pillars, was a large plate glass mirror. The old ranch parlour, my father said, was "a darn nice room." Throughout the house, there were a lot of books: Shakespeare, Thackeray, Scott, and other staples of British literature; a set of the world's great literature bought from a travelling salesman and hardly ever opened; a selection of Fabian books by William Morris, Bernard Shaw, and others; and a few religious books, among them an account of missionary work in Guinea, a birthday present to Joe from his mother.

Ranch house, driveway, and lawn, 1907. By this time the stump-strewn land around the ranch house had been remoulded and designed. Two concrete posts marked the upper end of the road to the lake, a wide gravelled driveway crossed in front of the house, and two levels of lawn were suitable for afternoon tea. Harris family collection, photo by R. H. Trueman

The small entrance hall was panelled with varnished tongue-and-groove cedar, and several rooms were wainscotted. Wallpaper was glued to the muslin tacked to plank walls. Ceilings were pressed tin. A variety of pictures hung on the walls: steel engravings of scenes from Shakespeare; the oil paintings of Joe's house at Westholme on Vancouver Island, by his half-sister Sophie; large, framed, and colour-tinted photographs of Joe's father and mother; a picture (which included Joe) of the Mill Hill cricket team; and a large framed reproduction of Henry Raeburn's *Boy and Rabbit*. There were a few Royal Crown Derby cups and saucers, and large curls of sticky brown paper suspended from the kitchen ceiling to catch flies. At the kitchen door were two large sinks where dirty hands, coming from the fields, were intended to be washed. The place was an enlarged farmhouse, but one with a certain elegance shaped by memories of lives elsewhere.

Life in this house became less formal over the years. Initially there were two dining rooms, one for family, the other for hired hands, a division that soon broke down. The family dining room was quickly abandoned except for parties. Meals were fairly formal, and always began with grace.

In the ranch house parlour, late 1930s. The fireplace and furnishings date from some thirty years before. From left: Margaret, my grandmother; Heather, her daughter; Joe, my grandfather; unidentified woman; and Ellen, my mother. Harris family photo

There was church on Sunday (Methodist for Joe, Presbyterian for Margaret) but absolutely no work; food was prepared the day before. One put on one's best, went to church, wrote letters, read, and enjoyed visitors if they came for tea—or so one did in theory. The young rebelled, the eleven external doors provided opportunity to slip away, and eventually Sunday sport was tolerated. Yet religion was in the air: one lived in God's creation in which, at least on the ranch, card playing, dancing, and drinking were not permitted (Dora Clemson smoked). Politics was in the air too. For Joe, the Slocan Valley was something of a promised land where civilization might renew itself, and in so doing rectify the wrongs of the decadent society that, in his view, he had left behind. Conversation was frequently political, and often turned to basic premises: that responsibility should override liberty, that society was composed of far too many parasites (Joe was particularly down on lawyers and smart wheeler-dealers), and that most good was accomplished by ordinary people doing the basic work on which society turned.

In sum, much had been put in place in a few years: a substantial farm, a still largely English way of life, a commitment to God and to social change.

The ranch was a complex creation, a product of hard work and innumerable transplanted ways, but also of a largely English critique of England. Yet it was situated within the British Empire, and the Harris family and most of the others who worked there thought of themselves as British, if increasingly, as the years passed, as Canadians as well. When Britain and Canada went to war, they too were at war, and it was their particular good fortune that none of the immediate family had to fight. My grandfather, well into his forties when World War I began, was too old. By a matter of months, my father was too young. In September 1918, then seventeen years old, he went to McGill University rather than the front. Had he lived, the older brother who died in childbirth would not have been so fortunate.

THE RECORD REVEALS NO MORE ABOUT THE BOSUN RANCH DURING ITS MOST expansive years. I can offer only these few supplementary inferences.

My grandfather poured energy, enthusiasm, and a good portion of a considerable English inheritance into his farm above Slocan Lake. He had pioneered once before, in the Cowichan Valley on Vancouver Island, but after an enormous amount of work there had sold and moved to the Slocan. There was hardly time, energy, or money for a third attempt. He was committed to the Slocan. He had the best farmland in the upper Slocan Valley, and before long a wife, children, and an expanding farm propelled by enthusiasm, hard work, and English money. Whatever Margaret's initial misgivings, the Bosun Ranch became her as well as her husband's Canadian home.

A substantial farm and farmhouse had been created remarkably quickly, almost as if the speculative energy that placed mines and pack horse trails high on Slocan mountainsides and concentrating mills and railways in narrow Slocan valleys had spilled over onto a narrow terrace above Slocan Lake. The speed and confident energy with which my grandfather committed himself to the Slocan, pulled up stakes in the Cowichan Valley, and began to develop his new property did not abate when he returned to the Slocan with a wife and set about turning a tiny clearing in a forest into a farm. He worked hard, hired labour, and the forest yielded.

In retrospect, however, thc whole undertaking was perilous. The modern Slocan was not yet ten years old and in flux; who knew what its future held? The Kerrs' assumption that there was an inexhaustible Slocan market for fruit and vegetables was no more reliable than promotional bluster about the size of Slocan ore bodies. Both, as it turned out, were wrong. My grandfather did not have the market he had assumed. Whether profitable commercial orcharding was possible in the Slocan remained an open question, but if it were, as he seems to have known as early as 1909, it would have to grow one or two varieties for an external market. The agonizing fact of the matter was that my grandfather had planted the wrong orchard. Short of pulling it up and starting again—a horrendous thing to do after all the work involved—there was no obvious solution.

And, for whatever reason, he built a large, unwieldy house. Perhaps he had in mind his family's home in Calne, and wanted to reproduce something of its scale and comfort. More likely, he wanted to please his wife. He had promised her an estate in Canada, after all. Perhaps the house was an atonement. Certainly, it was impractical—too many rooms, not all of them always used, too many external doors—and an enormous amount of work to clean, maintain, and heat. It was a full summer's work to haul, cut, and pile the furnace wood required each year; and a good part of each work week, year-round, for my grandmother, Dora Clemson, and whatever other help they had to clean and wash.

Overall, the ranch was sustained by my grandparents' commitment to it. The orchard was not pulled out, at least not yet, nor was the house simplified. After so much work and creation, and such pleasure in being there, the future was given the benefit of the doubt: perhaps the local economy would improve, perhaps ranch produce would find the local market for which it was intended. And so the ranch kept going—indeed these were its most vigorous years. By 1918 the ranch had acquired its own momentum, propelled by hard work, good times had there, pride in and affection for what had been quickly achieved, and an English inheritance that had not yet run out.

4 THE BOSUN MINE

IN THE AFTERMATH OF THE CALIFORNIA GOLD RUSH OF 1849, A PARTICULAR western North American version of industrial capitalism evolved in the hardrock mining camps of the American West. Well-established by the early 1880s not far south of British Columbia in the Coeur d'Alene district of Idaho, this hardrock mining complex readily crossed the border when, late that decade, silver-lead (galena) ores were found along Kootenay Lake. In 1891, it reached the Slocan. With it came more than a century's experience with industrial capitalism, and much particular experience with mining and shipping in isolated, mountainous terrain. With it, too, came the institutions, services, and equipment required to support these activities. Tucked-away valleys, virtually unknown to the outside world, could be transformed almost overnight into thriving mining camps because analogous developments were not far away.

In the Slocan, as in other mining camps, prospectors made the early finds, but the promising claims they staked were soon in the hands of companies that employed managers to develop mines and foremen to oversee miners working for wages. Telegraph lines, steamboats, and railways, all quickly introduced, established essential connections with the larger world. Machinery for mining and concentrating ore, almost all of it manufactured in the US, was available. So were the tensions inherent to industrial capitalism. With the miners came a militant union—the Western Federation of Miners (WFM)—formed out of labour strife in Butte, Montana, in the 1880s, hostile to the capitalist system, and involved in several protracted and violent strikes in American mining camps. From the mine owners' perspective, the WFM was a union of dynamiters and thugs; they formed mine owners' associations, and hired Pinkerton detective agents to infiltrate the union.

American miners and capital dominated the early years of Slocan mining, but the American presence diminished as more miners from eastern Canada and western Europe arrived and British capital began to invest in Slocan mines. Most of the principal American mining magnates turned down the Slocan, judging its small and fractured veins inadequate to support major mines. There was room for British investment, and the Slocan began to attract an edge of the vast flow of capital leaving Britain in the two decades before World War I. British capital and management, usually administered from London, were superimposed on an essentially American mining complex in a remote, mountainous valley thousands of miles away.

Such, briefly, is the background of the Bosun Mine discovered on my grandfather's property above Slocan Lake in the summer of 1898. It, like the ranch, was named after the Bosun, who had followed my grandfather from the Cowichan Valley to the Slocan. Throughout much of its erratic life, its owners were British. They managed a western North American silver-lead-zinc mine in the middle of my grandfather's farm that, in its own curious way, was also a product of industrial capitalism and of exported British capital. Side by side on the same narrow terrace overlooking Slocan Lake were two very different economies, values, and ways of life. I have considered the farm and its ways. Here I provide my grandfather's account of the discovery and early development of the Bosun Mine, and other, shorter accounts of its later development and management. I then offer some concluding comment.[1]

THE DISCOVERY OF THE BOSUN MINE FOLLOWED DIRECTLY FROM THE DISCOVERY of the Fidelity, a mine just off my grandfather's property to the northeast. Writing in the late 1940s, my grandfather described these beginnings as follows:

> AFTER I HAD bought my land I had had a most disagreeable shock. I found that my land was largely covered by two mineral claims that

belonged to the hotel keeper, Jim Delayney of the Grand Central Hotel [in] New Denver. This would prevent me from getting the title to my land and I had to buy him out. These claims were pure "wildcats."[2] Not a bit of work had been done on them. Delayney wanted $700.00, and although I wanted nothing whatever to do with mining I had to buy them.[3]

Then a strange thing happened. Some people in Silverton had "grubstaked" a young American named Frank Byron.[4] Byron had had no luck and was returning on the very last day of his grub-stake when he came on a tree that had been up rooted. There was some rock thrown up, Byron struck it with his pick and saw galena ore. Just a little picking and he had exposed a whole lot of ore—lead carbonate with galena. Byron staked the ledge, naming it the Fidelity in tribute to his faithfulness in staking it for his partners as well as himself. (He could have waited a day or two, and then staked it for himself alone.)

A little work showed up a very promising ledge with a fine showing of ore. We were working away clearing land when the excited Byron came across the clearing to tell us of his great discovery. He certainly was excited. It was evident to me at once that this discovery would have a very great effect on the value of my two claims.

Byron pointed out just about where the new strike was and the direction in which he thought it was heading. I knew that one of my claims must be very near.

This strike caused a great deal of excitement. Only a very little work showed up a good strong, though not large vein wonderfully well situated for cheap development. All sorts of people flocked to see it and all thought it looked very promising. Poor Frank Byron nearly went crazy with excitement. He and his partners put on a small crew and their first work turned out well. They had as much as nine inches of solid galena showing and the values in silver were very good.[5] A trail was made to connect with the New Denver trail and we all went up to look at the new mine. They built a cabin close to the workings and I supplied them with vegetables of course. There were about eight men there including Byron.

A man named Benedum, who was an assayer in Silverton, came up and experted the property. He made an estimate of the value of ore in the little hummock of mountain where the ore stuck out, if only the streak hung on. This talk sent Byron nearly crazy with excitement. His head swelled almost visibly. He was positively insulting to various mining men who tried to open up negotiations to buy the mine. "He had no time to talk with pikers." He hired horses and galloped about as if his time was tremendously valuable.

They borrowed money from the Bank and put on more men. I went up one evening with a few vegetables for the cook and went to see the showing of ore. It was 28″ across and practically solid galena in one place, but it was very much narrower a little lower down. They had put in two shots and were just about to shoot. I left as I had to bring up some more supplies next morning. Next morning the fine showing of ore was gone. There was only a streak of ore, from one to two inches wide. Such is mining in the small rich veins of the Slocan.

They worked with frantic haste and little judgement. They had ore that netted them about $80,000 but owed the bank $10,000 by spring. Then they leased the property to some good Cornish miners. These men did good work but had no luck. They sunk a pretty deep shaft but got hardly any ore and quit in disgust.

My grandfather, however, held two mining claims adjacent to the Fidelity, and was legally required to do development work. Although he had come to the Slocan to farm, these claims and the Fidelity discoveries drew him toward mining.

I GOT AN old fellow who lived at the Galena Farm and who was a bit of a surveyor to find out the exact boundaries of my claim, the

Tyro, and [when] we found that it was very near to the original strike of the Fidelity I hunted the mountain to find any indication of the vein. We found some rather likely looking quartz and very soon uncovered what looked like a vein. I did the assessment work, $200.00 for the two claims that I held on this showing. We got out a good deal of quartz but no ore, and the quartz tended to pinch out as we got further in. Altogether it was hard work for no encouragement. I got busy again with my garden and put in another very busy winter with the team.

When it was time to do another lot of assessment work, I got old Mr. Bartlett to make a careful survey of the workings of the Fidelity to determine the course of the vein as closely as possible. Bartlett did a very good job. We decided the direction of the vein pretty accurately, and where it should enter my claim as closely as possible. Then I got one of the best prospectors around, Ed Brennan, to work with me and put him in charge. He decided to make an open cross cut just below where the vein should enter my ground. We started at the very stake that Mr. Bartlett had set up. He worked east and I dug west. We dug an open ditch through the wash, which averaged about two feet deep. Just before the noon hour Brennan began to find "slick enside," that is rock that has been in the wall of a vein and still shows traces of the pressure and grinding that it has gone through. I had to go to town that afternoon but I hurried back and there was Brennan with several pieces of lead carbonates (decomposed galena). After a bit more hard work, we began to show up the ledge right in place, and next day we had a very nice little showing of galena ore. It certainly was exciting work and we both felt tremendously pleased.

C. T. Cross, the rather prominent business agent in Silverton, advised me to get the owners of the Fidelity to put a price on their property; then I would do the same for mine. We would put the property in his hands and then we could probably make a fairly good sale. We had done this and though I might have liked to try developing the prospect myself, I was bound by this agreement and decided not to ask Mr. Cross to modify it. My price was $7000.00 cash down [$7,500 in most accounts].

Mr. Sandiford, an elderly Englishman from Lancashire, had come out to look for mines for an English syndicate, the North West Mining Syndicate. He had been foreman of a mine in Serbia. A carpenter by trade, he was a shrewd and capable old boy, very self assured and boastful by nature. Sandiford came up to see the showing and almost at once wanted to know the price. I referred him to Mr. Cross and the deal went through.[6] We had dug two more holes in the wash farther down, and in each found the vein running strong and true with a little ore showing. Sandiford put men to work on the lower showing with the most astonishing results. The ore body got wider and wider the deeper they got. At ninety feet deep there was four feet of practically solid galena. "I knewed she was there," old Sandiford would say to all comers, and certainly plenty of people came up to see the bonanza.

Adit entrance, field level (level 4), Bosun Mine. This and other pictures from the *BC Mining Record* are the oldest visual record of the Bosun Mine, taken little more than a year after its initial development. *BC Mining Record*, Vol. 5, No. 8, Aug. 1899, p. 27

Initially, the Bosun Mine, owned in London and managed by W. H. Sandiford, was a huge success. My grandfather reported: "They took out an astonishing amount of ore for the little development required.

Sandiford opened up the mine in quite good style. In fact it was very simple to anyone who understood mining. They started at the lower levels, and as soon as the wagon road (less than half a mile) was built, began to ship fine galena ore in great quantities. The very first year they paid for the mine and a dividend to the shareholders."

The *British Columbia Mining Record* reported even more enthusiastically:

> ON JULY 4, active development began under the new company, and although employing only a few men at first, the force has been steadily increased until to-day [August 1899] it numbers over 30. Within two months of making the payment, and three of the commencement of operations, while still in the initial stage of development, shipments were begun, the first car leaving the mine [by barge on Slocan Lake] on September 6, to be followed by five others the same month, making a total of 120 tons for September, or a value of considerably more than the original cost of the property. From that time to this there has been no diminution in the productive capacity of the mine, the output being consistently maintained in the neighbourhood of 100 tons a month. In all, 9,000 tons have been shipped to date, having a net value of over $60,000, a record, I venture to assert, unprecedented in the history of mining in the Slocan and in all probability of the whole province.[7]

The North West Mining Syndicate built a wharf on the shore of Slocan and a short wagon road to the mine. The lake steamer, pushing a barge with several boxcars, would put in at the wharf, load the boxcars with ore, and take them, probably, to Slocan City, to which the CPR had recently opened a spur line. From there, Bosun ores went to smelters in Chicago; in Omaha, Nebraska; and even in Antwerp, Belgium. Many of the miners were put up in the two-storey log house my grandfather had just built.[8] He wrote:

THE OPENING OF the Bosun Mine gave a great lift to New Denver. The town seemed at last to have a definite payroll at the back of it. To facilitate the opening up of the mine, I let the miners board at the big log house. What rough manners they had and what an amount of spitting on the floor. They used foul language so habitually that they could hardly swear; they simply had nothing worse than ordinary talk to hurl at an enemy. They were a good hearted lot, full of fun, and loved to josh each other. Cleverley did the cooking and did very well indeed.[9] Even the miners did not kick, and they are desperately hard to please.

The initial success of the Bosun Mine attracted wide attention, not least for the quality of its management. The *British Columbia Mining Record* ended an article on the mine on this note:[10]

THIS ARTICLE WOULD be incomplete without a reference to the unique position in which the North West Mining Syndicate and its shareholders stand when compared with other companies similarly situated. Soon after the first annual meeting the directors were able to announce, with pardonable pride, the declaration of a 20 per cent dividend as a result practically of the first year's operations. This encouraging condition of affairs was due to several causes, a brief analysis of which might serve as an object lesson to our companies who have made a failure and then attributed it to the country.

To begin with, the company under consideration has an experienced London board, composed not merely of thorough business men, but those who have had to deal with mining ventures in other parts of the world, and therefore know precisely what they are about in this instance. Secondly, and of quite equal importance, is the fact that they are not burdened with a capital so vast as to be entirely unmanageable and out of all proportion to the scale of their operations. The third factor which contributed to their success is the unlimited confidence which they reposed

> in their local representative, Mr. W. H. Sandiford, who had full power to act for the company in any emergency which might arise. To his foresight and judgment, acquired during some twenty-five years' varied experience in every quarter of the globe, they owe a large measure of praise, and if there is one gratifying feature about the whole connection it is to know that his services to the country have been fully recognized by the directors and substantially acknowledged; a most excellent precedent for other companies who wish to achieve like success.

Less than a month later the Silverton Miners' Union closed the Bosun, demanding that men employed underground should be guaranteed a minimum of $3.50 for eight hours of work. This strike, following a provincial law that reduced the hours of underground work from ten to eight hours a day, and the mine owners' decision to reduce wages from $3.50 to $3.00 a day, shut down all the Slocan mines for the better part of a year. More serious in the longer run, the Bosun's richest ores were near the surface, and their zinc content increased with depth. The Bosun was becoming a zinc mine. Its ores were heavily penalized by North American smelters, while the cost of shipping to Belgium became prohibitive. Nor is it clear that the manager, W. H. Sandiford, was as effective as the London board and local mining writers assumed. My grandfather painted another picture:

> SANDIFORD WAS COMIC at first in his good fortune and very shortly tragic. He had been brought up a good Methodist, but now he quickly developed a drinking habit. His very boastful nature rendered him an easy mark for people who sponged upon him. He swallowed flattery as a cat goes for cream and could take it by the shovel full. Mrs. Sandiford was a very fine old lady, with far more sense than her husband ever had. It must have been terrible for her to watch him going to the dogs. Worse still, their son Charlie came out (very shortly after I had brought my wife).[11] Charlie was a fine,

Bosun Mine, 1903. Lower left: mine buildings at level 4. Above: dumps and mine buildings at levels 3 and 2. Photo by R. H. Trueman. Image G-00660 courtesy of the Royal BC Museum and Archives

> fresh looking boy, and we both took great liking to him when he first came up to visit us after his arrival.
>
> No young fellow could have had a better chance. He was well educated and very pleasing, with plenty of brains and not as boastful as his old dad. Both of the old Sandifords were tremendously proud of him, and he was made engineer to the mine and other properties which Sandiford was developing. The foolish old man started him off drinking also, and before long they were regular soaks. Men who wanted a job at the mine used to buy a bottle of whisky and give either of the Sandifords a drink or two and some flattery, and then almost certainly land the job unless the bookkeeper, Bob Thompson, took a hand, as he occasionally did, when the deal was too raw.

Whatever the accuracy of this assessment, Mr. Sandiford retained the confidence of the board in London. After the strike was settled, the Bosun reopened and remained in operation for most of the next three years, during which it was consistently among the Slocan's principal shippers. According, however, to the *British Columbia Mining Record*, by September 1902 the London board was considering closing it.

> THE THIRD ORDINARY general meeting of the Bosun Mines Limited was held in London last month. The Chairman, after paying a high tribute to the efficiency of the manager, Mr. Sandiford, explained that although the condition of the mine was excellent, it had been considered expedient to discontinue operations for a period, as the decline in lead and silver prices and the fact that the company was obliged to pay the miners a higher rate of wage, left little if any margin of profit. Recently, however, a considerable deposit of zinc-bearing rock had been encountered, and spelter [zinc alloy] being at present in great demand, it was hoped that the mine might soon again be worked. It was further stated that Mr. Sandiford had offered to accept a reduction of salary until operations were resumed, while the board agree for the meantime at least to waive their fees.[12]

A more definitive closing came late in 1903. The roughly 1,100 tons shipped from the Bosun that year was mostly zinc. In the hope of reducing shipping costs, the company had considered treating zinc ores at the mine, but had not succeeded in doing so. In these circumstances, the Bosun was unprofitable. There would be no shipments from it for the next fifteen years. Sandiford and his wife remained in the manager's house for a year, then moved to Victoria. (My grandfather said that Charlie drank himself to death and Sandiford "went insane and died in the Westminster Asylum still boasting of the Bosun mine.") A caretaker remained at the mine.

Little more than three years of work had produced a narrow industrial landscape composed largely of shattered rock from six adits (tunnels). It rose abruptly from a wharf at the lake, crossed a fairly level terrace with my grandfather's farm on either side, and carried on several hundred feet up a mountainside—the visible record of thousands of hours of underground work. Underground were six adits, one of them almost 1,300 feet long.[13]

Although the Bosun Mine was closed, the company directors were still considering how to handle its zinc ores. At its sixth ordinary general meeting, in London in September 1905, they reported and recommended as follows:

> OWING TO THE difficulties of mining conditions in British Columbia ... it has long been apparent that no favourable results from working the mine would be expected without smelting at the mine itself, or mechanically concentrating the ore before shipment. The cost of a smelting plant, and the fact that the supply from one mine alone could not be sufficient to work a separate smelting plant, renders the former idea impracticable. Your directors, therefore, have been giving much attention to the possibility of concentrating the ores. There have been difficulties, however, as to erecting a concentrating plant entirely for one mine, both as regards providing for the cost and sufficiency of ore supply; but an opportunity has arisen of considering amalgamation with a neighbouring mine which has a concentrating plant now in course of erection.
>
> As a result of somewhat prolonged negotiations, your directors have the pleasure to announce that they have concluded a provisional agreement with the Monitor and Ajax Fraction, Ltd., whereby in consideration of a certain number of shares in this latter company [26,000 ordinary shares of 1 pound each, and 6,666 deferred shares of 1 shilling each] the Bosun Mines, Ltd., will hand over its mines and property to the Monitor and Ajax Fraction, Ltd. By these means the Bosun and Monitor and Ajax interests will be worked together, and the shareholders of the Bosun Mines, Ltd., will obtain the benefit of the use of a concentrating plant.
>
> The mines and concentrating plant of the two concerns are immediately adjacent to each other, and can be worked as one

concern under the best conditions for economy, and your directors, believing this proposed arrangement to be in the best interests of the company, and as giving a probability of leading to good results, ask the general meeting to ratify the agreement.[14]

Rosebery–Surprise Mill, n.d, a mill financed in England, planned in London, and built in Rosebery on Slocan Lake. Intended to concentrate the Bosun's zinc ores, it never worked as expected, and eventually burned down. Image A006-002-009, Silverton Archives

The Monitor and Ajax Fractions were just above Three Forks, close to the CPR spur line from Nakusp, and some ten rail miles from Rosebery. The Bosun was four miles away by water. All their ores carried high proportions of zinc. With a closed mine on their hands, no other option on the table, and no other information available to them, the London shareholders ratified the sale. Work proceeded rapidly on a large concentrating mill in Rosebery, while at the Bosun's large bins were built at the wharf, and a gravity tramway built to the upper adits. Large volumes of zinc ore, some of it recovered from the mine dumps, could now be shipped to Rosebery. A year after the consolidation, the *British Columbia Mining Record* reported enthusiastically:

> THE MONITOR AND Ajax Fraction, Ltd., an English company operating in the Slocan district and owning the Monitor mine near Three Forks, and the Bosun mine near New Denver, and a recently completed concentrating mill at Rosebery, is planning to develop the mines named on a systematic scale. Both contain large quantities of zinc, which, under present conditions, is not favorable to active operations so the company is not mining just now. Certain improvements, found advisable, have been made to the

Bosun Mine loading bin, 1928. The rock in the foreground is from the adit just above the lake (level 6). The bin and trestle were built about 1905–6 to transport shipments of Bosun ores to the Rosebery concentrating mill. E. Wilmot/Library and Archives Canada/PA-014088

> concentrating plant, which is stated to be now working smoothly and making an average recovery of 95% of the lead 89% of the silver and 81% of the zinc content of the ore treated. During the two months it has been running when the district was visited lately, about 1,900 tons of ore had been put through the mill. Ore bins have been built on the waterfront of the Bosun property, Slocan Lake, and a gravity tramway to the mine constructed. This provision for shipping admits of the old dumps of ore being sent to the mill at Rosebery, for scows carrying four railway cars can be quickly loaded and the ore be thus cheaply conveyed to the mill. These shipping bins have a total capacity of about 400 tons and the cars can be filled from them in less than half an hour. Some development work was done last summer at the Bosun, this consisting of the extension of three adit tunnels, in each of which ore was encountered.[15]

Behind these plans were many untested assumptions, the most basic that the concentrating mill would work and be regularly supplied with

The ss *Slocan* pushing a barge loaded with freight cars, n.d. but probably 1920s. This was the link connecting the Bosun Mine to the concentrating mill in Rosebery or to the railway. Silvery Slocan Museum, New Denver, image 2001.014.001

ore. Neither proved valid. The mill did not solve the technical problem of separating zinc ores. The Monitor and Ajax Fractions closed, and in 1910 a forest fire destroyed their buildings. The Bosun did not reopen. Expected custom work did not materialize. A large concentrating mill in Rosebery stood unused and empty. Ordinary shares in the Monitor and Ajax Fraction Ltd. became virtually worthless.

Only well into World War I, and in response to high, wartime prices for silver, lead, and zinc, would the mine again become interesting. In November 1917, an American mining company, the Rosebery–Surprise Mining Company, acquired the Bosun, the Monitor–Ajax, and several other mines, as well as the concentrating mill in Rosebery. It improved the mill, and aggressively reopened the Bosun, employing some sixty-five men there through much of 1918.

The Rosebery–Surprise Mining Company operated the Bosun for a decade, during most of which the mine returned a modest profit. At any given time it employed some twenty to thirty men, and in most years shipped more than a thousand tons of ore, initially to the Rosebery concentrator where the problem of zinc ores was somewhat mitigated by high metal prices. In 1923, however, the problem was solved at the Sullivan concentrator in Kimberley, and two years later the Trail smelter, equipped with the new selective flotation technology, accepted all silver–lead–zinc ores without penalty. The Rosebery concentrator closed; Bosun ores were rerouted to Trail. At the mine, the upper workings were leased and their dumps gone over for discarded ore. The company worked the adit just above the lake, now almost 3,500 feet long, and below it, more than sixty-five feet below the surface of Slocan Lake, opened another level in 1927. These lower workings continued to show narrow veins of galena (silver–lead) ore.

In 1928 the Rosebery–Surprise Mining Company sold the Bosun Mine, which it must have considered largely worked out, to Colin Campbell, a businessman in New Denver. He managed the mine for a year, then, when a stope on the lowest adit was worked out, and with the prospect of another still lower adit beyond his means, turned the mine over to leasers and sought a buyer.

A report by the Minister of Mines described the mine:

> THE MINE HAS been opened up by six adit-tunnels and one level 100 feet below the No. 6 (or lowest) adit-level. This No. 7 level is approximately 75 feet below the surface of the lake and the vein has been drifted on for some 500 feet. During 1929 a small stope in the east end of No. 7 was mined by the owner and a total of 958 tons of milling-ore was shipped to the Trail smelter. The average grade of the ore shipped has been reported as: silver, 60 oz. to the ton; lead, 20 percent; zinc, 25 percent.
>
> The stope from which the above shipment was made played out, and in order to provide further ground for stoping operations it will be necessary to sink a shaft and start the development of a No. 8 level. This new development work has not been undertaken, efforts having been made to dispose of the property to a company which could handle the necessary development programme with considerably less risk than would be the case were the risk assumed by one man. Since August 1st, when C. J. Campbell discontinued mining operations, leasing parties are jigging the old dumps of the mine and making a product that has assayed 82 oz. silver, 22 per cent lead, and 23 per cent zinc when shipped to the Trail smelter. The remaining leasers are working underground, re-treating the old stope-fill and where possible mining any ground that is accessible and of a grade good enough to ship.[16]

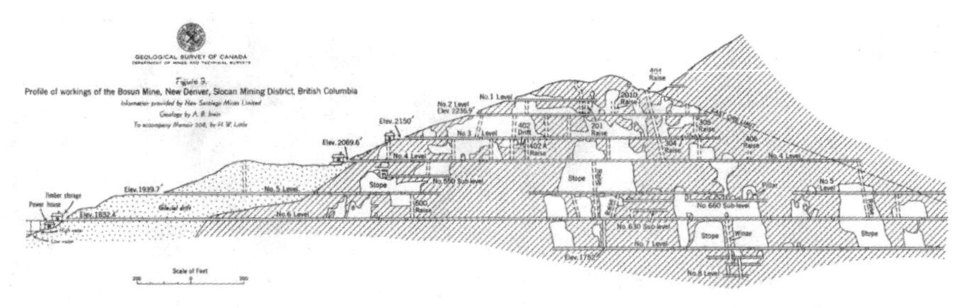

Profile of workings of the Bosun Mine, Geological Survey of Canada, probably 1930. Note the extent of the workings, the spaces from which ore was removed, and the two levels below level 6. Silvery Slocan Museum, New Denver

A year later, as metal prices collapsed and the Depression deepened, all Slocan mining companies laid off their men and closed down. A little leasing continued at the Bosun and elsewhere, but the primary economy of the Slocan Valley had virtually come to a halt.

WHEN THE BOSUN MINE CLOSED DOWN IN 1930, ITS ORE WAS ALMOST GONE, although leasers worked it intermittently for the next fifty years. In October 1956, there were reports that leasers had hit six to eight inches of clean ore running just over one hundred ounces of silver to the ton.[17] As late as the early 1980s, one of the dumps by the lake was being picked over, and sacks of ore shipped to Trail. When the highway (BC 6) that crossed an adit began to cave in, the Department of Highways poured in many loads of ready-mix and stopped up the entrances to the adits. The hole where a stope broke the surface was filled in. Trees began to grow on the dumps. The Bosun Mine had come and gone, its best years 1898 to 1899, its early life short, and a measure of order established only during a long decade after 1917.

The records reveal almost nothing of the miners or of their working conditions, though my father once said that there probably was so much salvageable ore on the dumps because angry miners chucked it there. On the other hand, the records provide a fair picture of the changing ownership, management, and productivity of the mine as capital reached aggressively into a corner of the Slocan Valley, there to create an industrial slice through the middle of my grandfather's farm.

The early mine was owned in England, managed indirectly by a board in London, and on the ground by a manager (Sandiford) appointed by the board. For all the vaunted experience and "precise knowledge" of the business and mining men who comprised the board, they knew next to nothing of British Columbia, nor of western North American hardrock mining, nor of the particular geology of the Bosun. Although the board was, in principle, the centre of calculation of the whole system, in fact it had little information to work with, and what it had came from the mine manager who, like most managers, couched reports to his superiors as favourably as possible. The board had little choice but to trust Sandiford; moreover he had recommended the purchase of the Bosun, and its initial development was spectacular.

Ore cars on the trestle above Slocan Lake, Bosun Mine, ca. 1960. Apart from occasional leasers, by the early 1960s the Bosun Mine had been closed for some thirty years, but its surface remnants were strewn along a strip of land stretching from the lakeshore, across the terrace that was my grandfather's farm, and some distance up a mountainside. Harris family photo

The Bosun's story, however, was common in the Slocan: rich ore, close to the surface and easily worked, then declining yields with depth. Nor was the board equipped to deal with silver ores with a high percentage of zinc, a smelting problem that, at the beginning of the twentieth century, was unresolved throughout North America. Probably Sandiford was neither as good as the board nor as inadequate as my grandfather thought he was; the problem was the mine, its ores attractive enough to draw speculative English investment, but not to sustain it.

The scheme to sell the Bosun to Monitor and Ajax Fraction Ltd. was even more risky. It made some abstract sense from afar: an integrated system comprising a concentrating mill built to handle zinc ores, and mines nearby known to be heavy zinc producers. But the mill hardly worked and the mines did not ship. Under English management, the Bosun was profitable for only a few initial months. Overall, the North West Mining Syndicate and Monitor and Ajax Fraction Ltd. were the means by which a good deal of English capital was transferred overseas and lost. Its residue was in the effects of wages paid and goods purchased, in idle mine workings, and in a large, empty, concentrating mill in Rosebery.

The Rosebery–Surprise Mining Company that purchased the Bosun in 1917 may have been better managed—it is impossible to know from the records at hand—and certainly benefited from higher metal prices and accessible smelting. Prices for silver, lead, and zinc, pushed up by the war, remained relatively high through the 1920s. Moreover, the problem of zinc ores was solved by 1925, when such ore could be shipped to Trail, not far away. There seems a certain shrewdness in the Rosebery–Surprise operation, particularly the decision to sell in 1928 when prices were robust and the mine appeared to be nearing the end of its life. In fact, by 1928 when C. J. Campbell bought the mine, the Bosun's years were largely over, less because of the Depression, decisive as it was for a time, than because there was little left to mine. In a span of just over thirty years, during barely half of which the Bosun was an active mine, industrial capital had plucked a valuable resource, removing it permanently from the valley.

Richard White, a prominent American environmental historian, has argued that a mania of late-nineteenth-century railroad construction in the American West resulted in far more miles of track than were needed; in fortunes for a few railroad barons, often at the expense of the companies they ran; in abnormally low commodity prices (because of the glut of primary resources coming on market); and in settlements on difficult land where pioneers faced only hardship.[18] He holds that there were alternatives, and that the American West did not have to be developed as rapidly as it was. There was an opportunity, he suggests, for a very different outcome.

Such a claim had long preoccupied my grandfather. In general, he thought different outcomes were possible, but only if cooperation replaced greed, and if wise, centralized planning replaced reckless speculation (see chapter 6). He thought British Columbia had built far too many miles of railroad, and had surveyed and opened up far too much land for settlement before older areas were adequately populated and developed. He advocated the public ownership of most of the means of production. But he was a booster of mines. He had long argued that undeveloped mining claims should be taxed more, and developed claims less. Moreover, for purposes of taxation, owners should assign their claims an assessed value, after which anyone willing to pay this price plus 5 percent could purchase them.

In these ways, undeveloped claims held as speculations—the curse, as he saw it, of any mining district—would come on the market, and low taxes would encourage their development. As a farmer who had intended to supply the mines, it was in his interest to support them. At the same time, he believed in the public ownership of the means of production and, partly by this means, in a cooperative society. Yet in all his political writings I find no direct reference to the nationalization of mines. Perhaps he recognized how difficult this would be, partly because the whole ethos of the speculative mining rush that had developed the modern Slocan contradicted his managed, cooperative vision, partly because the small, rich, uncertain Slocan mines, of which the Bosun is a good example, were particularly poor candidates for nationalization. Moreover, in the 1930s, when these issues came most insistently to the fore, all the Slocan mines were closed.

5 THE BOSUN RANCH, 1924

SOMETIME IN THE LATE FALL OF 1923 OR EARLY IN 1924 MY FATHER, DICK Harris, returned to the Slocan from Montreal where, as a graduate student in English at McGill, he had been working on a master's thesis on George Bernard Shaw. He settled into the ranch house with his parents and siblings, and while finishing the thesis immediately entered into the work of the farm and the social life of the valley. He had his eye out for a teaching position in the New Denver high school but, not having the requisite credentials, would leave the ranch at the end of the summer for Vancouver and the Normal School at the new University of British Columbia (UBC). For most of his months at the ranch he kept a diary: three small, black booklets in his legible, grammatical hand.

These diaries intersect with the ranch at a pivotal time. The orchard that had been intended to provide its income had failed. Another economy would have to be found—dairying was the principal candidate—but in 1924 the ranch was little more than a subsistence farm. This economic predicament pervades the diaries. Moreover, decisions were being made about staying or leaving, decisions that particularly affected four people at the heart of these diaries: my father, his younger brother Sandy, and their friends Mollie Colgrave and Gertrude Smith. Mollie had grown up in Edgewood on the Arrow Lakes, the daughter of an English family lured by spurious orchard promotions. In 1924 she worked in the telephone office in New Denver. Gertrude, the daughter of a Scottish general store keeper and his wife from Northern Ireland, had returned to New Denver after an undergraduate degree at UBC, and worked in her father's store. Mollie and

J. B. Smith's store (centre) and the Newmarket Hotel (right), n.d. but probably mid-1920s. During the months of my father's diary, Gertrude Smith worked in her father's store. Silvery Slocan Museum, New Denver

Sandy would stay; Dad and Gertrude would leave, she to graduate work in zoology at UBC and Berkeley, and he, after a year at the UBC Normal School, to high school teaching in Vancouver. The next generation on the ranch was being sorted out.

In the first of the two parts that follow, I provide selections from my father's diary, a third, perhaps, of what he wrote. My selections are intended to reveal the ranch economy, the local off-farm activities in which the ranch was situated, and the character of my father. The principal casualties of these criteria are my father's observations about provincial politics, church services and ministers, and a few of his friends. Then briefly, in a second part, I reflect on the ranch as it was in 1924, on its ideological underpinnings, and on its rapidly changing social relations.

The selections below are the complete entries for the days indicated, unless ellipses indicate omissions.

RICHARD COLEBROOK HARRIS, DIARY, BOSUN RANCH, FEBRUARY–JULY 1924

Feb. 1 Still mild. Worked all morning hauling out mining timber while Sandy felled. In the afternoon we two hauled 351 feet to the Bosun. Dad reported no more wanted for the present which is rather a blow. Very little rain and in the afternoon a good deal of bright sunshine—Like a day in spring. Hockey prospects poor. Spent the evening for the most part reading "Black Oxen" by Gertrude Atherton.

Feb. 3 A quarantine Sunday. Finished "Black Oxen" this morning. A clever book, sensational but with a well-thought-out plot and a logical, true to life conclusion. I can't say that I took very much to either Clavering or the Countess Zattiany or, for that matter, any of the "sophisticates." The picture of the ultra-civilized (self-styled), semi-bohemian, aristocratic and literary people of New York augurs ill for our civilization if it is a true representation of it, and I can quite believe that it is. Wrote to Kath and did a little work.

Feb. 4 Sandy and I decided on a logging trail first thing this morning. Most of the morning spent in fixing up the trail and getting one load over it. Three small loads only eight small logs in all were the total for our first real day's work but we are in a position to go ahead faster now. Prospects look, on the whole, good, though there may be considerable loss from rotten cores. I have started to keep a time account and an expense account of the whole logging job. Spent the evening vamping and practicing songs and reading over parts of my thesis. It seemed to read well tonight and may not be so bad after all. But I must get busy with it if I am to finish.

Feb. 7 Dad and I made good headway this morning. The block and tackle worked wonders. In the afternoon out logging and were surprised by Gertrude and Mollie. They came in to tea and we quite enjoyed ourselves. Such a relief to see someone who is not a Harris. Played the piano a bit at night and did a bit of work.

Feb. 10 Miserable wet morning in bush. In spite of it being like spring we went to Slocan for hockey. There sadly disillusioned for after all our big talk Slocan beat us 9 to 8 after 10 minutes overtime. Ice heavy with water lying in parts. Trying to tempers on both sides. The Denver team seems disorganized and at times resorted to foolish tactics. Tired on boat afterwards and managed to get a fairly good sleep.

Feb. 11 Home from Slocan about noon. Humiliating to have to explain our defeat. In the afternoon Dad and I rustled some furnace wood while Sandy worked on the car engine. Went to bed very early.

Feb. 15 Hauled manure this morning; reflected the while on my occupation and contrasted it with my life in Montreal. What would some of my college friends think to see me at it? Hauled a good log after dinner and rolled it into the lake. Rode down on Mickey [a horse] at five, had a haircut and saw Harry Angell. After supper got ready to go as "Knight of the Bath" to the Silverton Masquerade. Felt rather out of place at the masq.; how few there are with whom I seem able to have real companionship! Nevertheless had rather good fun out of it. Heather [sister] and Joe [i.e. Tod, brother] and I walked home afterwards. Bed at quarter to four.

Feb. 18 Rather a poor day at the logs. The lack of snow is making it a very hard job. However, it is progressing slowly. We hear that the price of a new Chevrolet f.o.b. at Vancouver is $895, tax not included. This is a frightfully high price. Canada certainly is an expensive place to live in. It rather puts a crimp on our new car aspirations. It means that by the time the tax is paid and the car brought here it will cost very close on $1000 which is far too much for a Chevrolet quoted at $490 in the States.

Feb. 20 ... read some of Heather's compositions tonight. She writes quite well and displays a quite original mind and a certain gift of expression. Her punctuation might be improved. It is a pity she is such a poor speller.

Feb. 21 Took the afternoon off and went down with Sandy to register. We are now duly qualified electors of British Columbia. It was quite an experience going downtown. Perhaps we have been sticking a bit too closely to ourselves lately and not getting out about enough. Mollie and Gertrude walked out with us and we had a very enjoyable musical evening. The walk in and out again afterwards was a bit too much. To bed tired at 2 A.M.

Feb. 23 Have felt the effects of two successive late nights today. Almost beyond question late nights are something to be avoided. What fun one has seems to be more than offset by the discomfort next day ...

Feb. 26 Did several odd jobs about the house this morning. Hauled over old sleigh for storage, fixed up smoke house etc. This afternoon hauled down three good loads. Bruce Campbell brought up a car this morning. Did a good deal of work on the thesis tonight and read over the chapter on Shaw's social political and economic theories to Dad and Mother. They seemed to think it was quite good, which is encouraging.

Feb. 28 Sandy and I stamped and scaled the logs at the lake this morning. We had about 7000′ there. This is as much or more than we expected but it seems slow work on the whole. However we are making good wages at it. Mother, the two Mollies and Gertrude showed up on the flat just as we felled a tree this afternoon. Too tired to do much work tonight. Have decided to turn down the dance tomorrow night in Silverton. I simply can't afford the time and the late nights are too much.

Mar. 1 Had the car running for the first time today. We had to haul her about for some time with the team to work the stiffness out of her bearings. She runs only fairly well but it is a great blessing to have some means of transport. Trickett wants me to speak at the P.S.A. a fortnight hence ...

Mar. 3 I enjoyed quite the self-satisfied feeling that one has after a good day's work this evening. We got down about 1750 feet today which, considering that there was nothing cut ready for us, is about our best day. Had my first installment of the thesis from Miss Osborne [typist]. She seems to have done a very good job. I wish the duplicate copy were on the same quality paper; otherwise it seems to be perfectly O.K. Helped Heather with some Latin ...

Mar. 6 Quite a fair day in the bush; in fact so far this week we have averaged well over a thousand a day so we are quite satisfied. Tonight I had a good opportunity to get some work done and did do quite a lot but it was broken in upon by helping Heather with her arithmetic in preparation for an exam tomorrow.

Mar. 8 Sported my new riding breeches today and I am very well pleased with them. They seem too good to be out working in. Pincher was very lame this morning so we couldn't use the team all day. There are now four trees down and ready to come out. I sincerely hope that he (Pincher) will be quite alright again by Monday but it looks doubtful. Joe and I worked up the hill together all day. Worked a bit this evening but as usual was very sleepy.

Mar. 9 Raked off the lawn this morning and finished up some work on the thesis. I can now send Miss Osborne three chapters. I certainly shall be glad when this thing is off my mind ... Missed church tonight because of a puncture [presumably in a tire of the ranch truck]. Spent quite an enjoyable evening and then when taking Mollie and Gertrude home got another puncture.

Mar. 10 ... Sandy brought home the mail late and there was a letter from Kath. It certainly is a pleasure to have a good letter and a good letter Kath certainly can write. She surely has brain. I am feeling quite tickled and bucked up and stimulated and everything else. Kath quotes Shaw like a devotee and I don't really believe she thinks much of him. She surely remembers what she has read and is agile at argument.

Mar. 11 Rather a poor day. Not enough sleep last night served to keep me cranky and on edge all day. Nothing knocks me out more than the loss of a little sleep unless possibly it be over eating or foolish eating. If I sleep regularly and for about eight hours and am reasonably careful with my feeding I seem able to keep at all times as fit as a fiddle. But in spite of knowing this I still violate the rules frequently. Tel est la vie! Went with Dad to the church annual meeting tonight. I was elected chairman of the board of managers which board is, I believe, responsible for the church finances ... I am also on the executive of the music committee and that will take some time.

Mar. 12 Wrote less than a page of my thesis tonight. I'll be caught at this yet if I don't get right after it. I phoned to Mr. Trickett that the subject of my address would be “Some Impressions of Montreal.” I hope I can make it interesting. I must begin to put a little work on it.

Mar. 13 Another evening wasted so far as my thesis is concerned. This will be getting desperate soon. There is the play and the rehearsals further to complicate matters. I shall certainly have to take some time off to it. Tonight we have at least settled on a play. Gertrude and Mollie stayed to supper and then we read through about five plays. Finally decided on “Rooms to Let” which hasn’t much in it but which I think is the most suitable to finish off an evening with. As yet the cast is not yet settled upon. That is left to me and some job it will be.

Mar. 15 ... in the afternoon we had our first ball game of the season when we beat the intermediates 19-15 in a quite decent game ...

Mar. 17 This thesis hanging over my head will drive me crazy. I must finish it right away. Another night gone and not a tap done on it. Then there is this P.S.A. thing to prepare for, so there can’t be much done this week. I simply must take some time off. Tonight we made final arrangements about the cast for the play. Mrs. Stevenson is to be Mrs. Smythe. I rode in on Joe’s bike which was quite enjoyable for a change.

Mar. 18 Stayed in this afternoon partly because I wasn’t quite up to the mark and partly because I wanted to do some work on the thesis. I got something done but not a great deal. My wretched tooth which bothered me last fall in Montreal has started to ache again. I took my first aspirin tablet tonight. It certainly relieved the pain but I distrust them. Its effect on my heart action was quite noticeable.

Mar. 20 Stayed in and wrote this afternoon but did not accomplish quite so much as hoped. Mollie and Gertrude came up and so I didn't do anything this evening. We had a very pleasant evening sitting around the parlor fire talking and fooling. I must bestir myself about some sort of an address for Sunday right away. If there is going to be any sort of a turnout I should try to have something half decent.

Mar. 21 Spent the whole day on the thesis and made some progress. I have still two chapters to write but I think that I shall make them short. Tried to shoe Boy today but made rather a mess of it and gave it up in disgust ...

Mar. 22 ... Stayed home from the first rehearsal for the concert tonight to get something together for the P.S.A. tomorrow. I think I have enough to keep going for twenty minutes or so. I shall have to be careful to watch the time and not wander on for about ¾ of an hour.

Mar. 23 Church this morning. Mr. Stevenson was in quite good form but I am beginning to get more and more skeptical of the value of preaching or rather of its effectiveness. Rather disgusted with my talk at the P.S.A. I was too long for one thing. Then I brought in a lot of stuff which it was almost impossible to make interesting. I think if I was doing it again I could make a much better thing of it. The heat of the room also helped to put me off. Mr. Nelson, Mollie, Gertrude, Ev. all up for supper. We had some quite enjoyable music afterwards.

Mar. 26 I suddenly realized that I have only 4 days to do practically two chapters and that Monday is positively the latest that I can leave the thesis. This means that I have got to have at least 2 days off. Today I went to Dr. Carter. I do not believe that it will be a very expensive business to fix me up. Somehow he does not appeal to me as being a very good dentist ...

Mar. 27 Spent the whole day inside. Progress was slow but there was at least some progress. I simply must work towards getting my brain to function more quickly and accurately. I am sure it can be done by keeping one's wits about him and one's mind closely riveted to the subject under attention. Poor old Frank Edward who fell over the cliff at the Apex yesterday was found all broken up – legs, arms and neck all broken and head bashed in. This will give the Sandon road a bad jolt I'm afraid. The Canadiens won the world's championship today. They surely must be some team. There has been some more political scandal this time connected with the Ontario government. It looks like a bad business. What with the Teapot Dome Oil scandal, the Squires affair in Newfoundland, and now this it almost looks as if our materialistic, money-grubbing age is beginning to raise Cain with our public men.

Mar. 29 Sandy and Joe made excellent progress today, taking out the last log from up the hill. They must have done over 2000 feet, which is the record for the job in one day. There now remains to skid the logs at the bottom of the chute, to scale, and then to haul to the lake. There must be close on 2 weeks hauling for a team and man as there are over 20,000 feet by the boarding house. It is a relief to see the end of the job approaching. It is also a relief to have the end of my thesis in sight. Today I finished the chapter on religion and just made a start on the chapter dealing with Shaw's technique. It will be a short chapter but I'm afraid a difficult one. The winter showed that it still had a bit of kick in it. This morning the ground was covered with about 4 inches of snow and a cold north wind blew most of the day.

April 1 The scaler was up today. I have not so far totaled up the amount but I think it is just over 30,000 ft. Went to practice tonight whence I got home at 11 o'clock. Then settled down to finish my thesis, which I did at 3:10 a.m.

April 2 Today I sent off the last of the thesis for which may the Lord be truly thankful (!) Picked up rocks all day. Rehearsed the play once this evening. Drove GooGoo out afterwards, had a cup of coffee and then drove her back. GooGoo is nice to talk to for a change but she is too much among the gossipers down town methinks. Lord but she can say some cutting things when she wants to.

April 3 This has been what I call a wasted, futile day. I picked rocks this morning – a job requiring certainly very few brains, but what I like worse about it is the feeling of uselessness that there is about it. You are doing this for a hay crop, which even if good gets you where? What is all this Dairy business bringing in? It seems to me that here, we are not concerned nearly enough about our actual products. We are not particular enough about the milk after it gets to the Dairy and of what becomes of it all. If we are not producing stuff and selling it in pretty good quantities what is the use of all the work. From now on we have got to have an eye more to the returns before we begin any job. I saw four pretty deer in the Bosun field today. That grazed and ran about quite close to me. Although their tails were white I think they were mule deer. This afternoon practically wasted. I helped Sandy ... practiced a little baseball, and then hurried home only to have to go to one of these wretched rehearsals which was rotten.

April 4 Spent the day picking rocks. This evening I stayed home while the rest went to see "The Town which forgot God." I had a quiet but pleasant evening – cleaned up, candled the eggs, and then practiced my part in the play a bit. Had dropped off to sleep in the kitchen before the others came home.

April 5 It rained heavily today so this morning I churned and made up nearly 19 lbs of butter. Tried a little fishing with Joe after lunch but with no luck. Pruned for a while this afternoon. This evening we all went to the band social which was quite a success. A good friendly sociable spirit was always in evidence.

April 6 Heard a splendid sermon from Stevenson today except for a few ideas at the end of it. The idea that it is possible for a man to be brought to a swift and horrible death simply because he had said "I shall put aside Christ for the time" is not worthy of a man with the thought of Stevenson. After church we went to Smith's and after I got home I began a letter to Kath.

April 7 Finished picking rocks off the Far Field this morning – a thankless job if ever there was one. This afternoon I was engaged in a still more melancholy task i.e. pulling out fruit trees. It seems a terrible thing to be doing – so much effort and hope in the past seems to be wasted but it is almost necessary. Some of them are diseased and they are nearly all trees which will pay back very little. The concert tonight went quite well but for the play, which I thought went poorly. We must improve it quickly.

April 8 Made a small start on plowing in the orchard today. The rehearsal tonight was rotten. What with people away and fooling about it was very little use at all. We have simply got to get our parts for the play learnt by next practice and start to take it a little more seriously.

April 17 Ploughed and harrowed today. Tonight we had quite a good rehearsal of everything, the play included. The concert seems to be rounding into shape and, I think, will be a big success.

April 19 Finished the ploughing etc. for a time at least and then started to snake out some more mining timber. This afternoon Sandy and I played our first tennis of the season and did pretty rottenly. Rehearsed the play three times tonight. It is gradually getting to be fairly well finished and a good many of the ragged spots are rounding into shape.

April 20 Fine Easter service this morning. The choir didn't do badly although some of us, especially the "basso profundos," hit some wrong notes. I slept badly last night on account of my tooth. This morning my face was swollen and it has continued to swell all day. Stayed home from church tonight (Knox Wright preaching) and to bed early.

April 21 Felt too miserable this morning to do much work. The second ball game between the band and the town was played today. The town won 15–5. It was rather a slow uninteresting game. My arm nearly played out before the end of the 9 innings. It was good practice at any rate ... The rehearsal tonight wasn't very good; two absent. The play didn't seem to be all that it should be either. There was a long discussion after it as to whether we should go to Silverton to the dance. I'm afraid I killed it but I can't help but feel glad. It would have raised the dickens with us, especially with this wretched tooth. I feel a bit of a piker though and I shall have to make amends sometime in the future.

April 22 Worked rottenly today; this concert is becoming a curse. Tooth slowly improving. Hoped to be down early today and attend to a number of things, but owing to the car being out of whack we were almost late after we had finished supper at Harry

Angell's. Semi-dress rehearsal seemed to me to be awfully flat. Mrs. Campbell wants me to make up a verse for her song.

April 23 I had one of the worst toothaches after dinner today I ever had in my life. The dress rehearsal tonight was discouraging. Fred and Stedman away balled things up badly. It may go well tomorrow night but I hae my doots.

April 24 I had the most painful session with doc. Carter today that I ever had I believe with any dentist. He at any rate relieved me. The concert went off swimmingly – hardly a hitch. Everybody immensely pleased. A sort of reception at McFaddens' afterwards – very enjoyable.

April 25 Another session with Doc Carter, not so painful this time. The town were badly beaten in today's ball game against the band. I couldn't pitch up to the mark, and our fielding was rotten.

April 29 Sandy and I got another good day at Bosun mining timber this afternoon. I delivered $34.40 which we had got ready in the morning. Tonight we rehearsed the play which also went as flat as a pancake ...

April 30 Hauled stall timber to the Bosun and finished our work there for the time being. Our cheque this month should be about $140.00 ...

May 1 Started hauling our pine logs to the lake and made 8 trips taking down possibly 2000 feet. There should be no difficulty maintaining this rate throughout the job and even at that it will take two weeks. I was at home this evening for a change. Started to read "Fanny's First Play" to Mother. Heard from town that Mr. Irwin and Ivy could not come to Sandon tomorrow, which is a thousand pities. Nevertheless the rest of us are going. Gertrude is going to take Ivy's place in the play. To bed early.

May 2 Hauled logs this morning. Stopped work at noon. We were a jolly lively party that left N.D. for Sandon ... The show itself was not quite up to the mark but the Sandon people were on the whole pleased. The dance after it was fun. Miss McLeod the nurse and "Olga" Swan are surely some wenches. They are what Wally Willard would call "hot babies."

May 4 Broke the Sabbath once more making butter. Mollie came up after Sunday School. She is like a little ray of sunshine; always full of fun and good spirits. Walked in to church with Mr. Nelson and Dad. Dad got quite warmed up in talking to Charlie about his social theories. Dad's idea that the parasites must be eliminated from society is a good one and it certainly is right. But how? His sweeping suggestions seem to me to be almost impossible of realization. It is all very well to talk of pushing out the idlers and the useless busybodies and cutting off their supplies if they will not work but who is to do this? Can we expect this to be done well by the average government officials as we know them? It seems to me that a long process of education is required (perhaps Dad's work is a step in that process, who knows) and that the desired end is more likely to be brought about by a slow and more gradual socialization or integration of humanity. I am inclined to believe more in Mr. Sidney Webb's "Inevitability of Gradualness" provided of course there are those at the head who always keep the ideal in sight and who see that as a touchstone to apply to every measure or reform which comes up. There was a good service tonight, but only 3 of us in the choir.

May 5 Sandy left for Nelson this morning to get the new car. I managed to get down five good loads this morning in spite of a fairly late start. This afternoon I moved the Mears, bag and baggage, into town in three quick trips with the truck. They are not very good tenants and in some ways I am not sorry to see them go ...

May 6 Hauled 4 loads this morning. This afternoon I fixed up the dolly and with the help of the Bosun blacksmiths I think that I now have it in first class shape. Pat Henry phoned over the news that a calf was in the Bosun shaft. We hurried over with Pincher, rope, pulleys, etc. fixed up a gin pole and lowered Bruce Campbell into the shaft. He reported the calf to be standing, and with a broken horn as the only apparent injury. If the animal has not some serious internal injuries and survives it will be a miracle. On account of the wind and rain and dark we were forced to leave him there for the night. He apparently cannot fall any farther. We threw down some hay to keep him going. Talked with Sandy at 10:30. He has bought a Chev, and seems highly pleased with her performance. I am rather looking forward to seeing her.

May 7 Raised the calf first thing. He is thin and weak and sore but I think will be alright. Ran Dad and Mother down to the boat. Dad will be away down the valley till Friday organizing for the Provincial Party. Mother went to Slocan to meet Sandy and the car. The car is a little beauty ...

May 16 Joe Gautier phoned over at noon that our logs were drifting away which news gave me at least a sort of sick fainting feeling for a while. We rushed down and put a temporary boom around them. Sandy got Marshall's launch to round up the only one which apparently was adrift and this proved not to be ours. The boom is only a light temporary one so I hope Lengle and Johnson will hurry up with theirs. We appeared in our concert positively for the last time, at Silverton tonight. We were fairly well appreciated though not so well as in New Denver. It was good fun at any rate.

May 17 While at work on some new wheels for the dolly we heard Greenwood's big whistle at the Bosun wharf. She brought our boom, much to our relief. After supper we worked till dark and skidded most of the logs into the lake.

May 18 Finished the logs this morning and fixed up the boom. They were gone by noon, as nice and compact a little boom as one could wish to see. My only regret is that it is not several times as large ...

May 19 Sandy and I stacked up a cord of logging today and we worked hard to do it. One would never make a fortune at logging. Tonight I went down to a committee meeting to make arrangements for the 24th of May. We drew up quite a program of kids' sports. We have over $80 for prizes so we should do fairly well.

May 20 Planted potatoes all day. The three of us got in 6 sacks which I consider quite satisfactory ...

May 21 Harrowed the potatoes, pulled more fruit trees, and hauled manure. Our orchard is all shot to pieces. The way the trees have died off is the very limit. There must be at least fifty more ready to come out.

May 22 Spent the morning dung hallin'. This afternoon I went to see Dr. Carter for the last time I hope. He gave me my bill, $47.00. I nearly gave up the ghost. And I already owe Miss Osborne close on $40. Truly there is no rest for the wicked. We had a practice game against the kids today, which was good fun and a good warm-up for all of us.

May 26 Finished hauling manure and started to plough again. I had a letter from Muriel [Aylard] today in which she referred to several nice things she had heard about my thesis. It certainly was encouraging. Dad started a straight milk diet today. He has to take 2 gallons every 24 hours (the equivalent in food value of 96 eggs I am told). So far he appears not to be suffering any bad effects.

May 28 Ploughed and harrowed all day. I heard just before supper (Harry Angell phoned up the news) that I was an M.A. I am relieved and thankful that it has come out alright but it is surprising how little the news really affects me.

May 30 Spent most of the morning making butter. This afternoon started on the housework. Sandy has got a job on the Slocan road and will leave on Monday. That will mean that Dad and I are going to be very busy. The remainder of the pine logs, Bosun logging, our own wood, ploughing, are going to keep our hands full to say nothing of haying, gardening etc. etc. However I think on the whole that this is the best policy. Just at present there is not enough *paying* work for the three of us ...

June 6 ... Took the money at the show tonight. The picture was "My American Wife." I like Gloria Swanson very much. There is something awfully attractive about her I think. I had a long letter from Kath today which, as usual, was very interesting.

June 8 Drove Dad, Heather, and Joe down to Sunday School and then went for a run to Silverton with GooGoo. My conscience rather pricked me for this, but when one thinks of it, it is really quite a small extravagance. Harry Angell, Harry Avison and Mollie came to lunch. We went down for Gertrude later and ran on to Silverton on the way back (more extravagance) ...

June 13 Dad at Rosebery today, hauling up the S.S. Slocan on drydocks. I made butter etc. etc. Three people brought up cars today.

The dance tonight after the picture show, I thought went flat. What was worse I didn't collect any money for it as I should have done. I hate collecting money like that and I suppose I shirked it. Now I suppose that the dishes have to be paid for out of nothing. The rest seemed to enjoy the dance.

June 19 Drove Heather and Joe to school this morning and when I returned home I lubricated the car completely which took most of the morning. Too wet to work outside this afternoon so Dad and I stayed in. I did not do much as I had the beginnings of a bilious headache. The picture show tonight was a good one, "The Spanish Dancers." I had seen it before in Montreal. Took in $26.50 tonight. Had a long talk with Sandy before going to bed.

June 21 In answer to Mr. King's request I went down and gave the matric class a little talk on the Romantic revival. I hope it was of some use to them but I really think it helped me more than them. I quite enjoyed the brief review. This afternoon Joe and I hauled two big loads of logs and we are all set to go ahead on Monday ...

June 30 This morning we cocked hay. In the afternoon Joe and I fixed up the hayfork rigging and cleared up and put in place the road over to the Bosun. We hauled our first load of hay after supper working till 9:30. Sandy was down all day.

July 5 Finished the hay on the Far Field today. The crop is very light. Joe, Heather, and I went down to tennis tea this afternoon. I had three good sets ...

July 8 Cut the field at the back today. It is a splendid crop, in fact the only heavy hay crop on the place.

July 11 Sandy came down today. Mollie, Fred and he went to Sandon tonight to play at the Orangemen's dance. They made a hit there apparently; many described it as the best dance and the best music ever in Sandon.

July 12 Had another slight but annoying bilious headache today. Drove Sandy and Fred over to Rosebery from whence they depart for the fires.

July 15 Hauled carbide to the Bosun. To have an excuse for calling at Hoben's store I ordered a rice pudding dish there, and then did not have the nerve to broach Hoben with the question I wanted to. I wanted to consult him about applying for the N.D. High School job. However I made an appointment with him for 7:30 and we got on well. He advises me to write for a permit.

July 16 This morning I wrote to the Dept. of Education and with Mrs. Hebron's help I think I got a fair letter drafted. I have my doubts about it, but should they by any chance grant a certificate then I believe that everything will be lovely. The magnitude of the job I may be letting myself in for sometimes appalls me. Halcyon's community choir got off to a good start tonight: 18 at practice.

July 18 Wrote to DeLong, inspector of High Schools this morning asking him to approve me as a High S. teacher. Put a shoe on Boy. This afternoon I ploughed the orchard. The show tonight was quite fair, though the story was full of improbabilities.

July 21 It rained today and I did not get much done. I churned again however. We shall now have to churn twice a week and I expect that there will [be] between 30 and 40 pounds. Joe and I did some work on both cars. It is difficult to say just what can be the matter with the truck. Gertrude and Aileen walked out to get something or other and Mother and I walked all the way to town with them and back. A letter from Mrs. Twigg [formerly Nancy Hill, Dad's Latin teacher on the ranch] leads me to believe that there is some chance of my getting the Creston [teaching] job.

July 22 Put in a good day today. This morning I ploughed and this afternoon I spent teaming down town. I hauled loads for McPherson's, Angell's and Clyde White ...

July 23 Hauled Checkalero's sand today and also some furniture etc. for Eccles. Every little [bit] helps. Tomorrow the team goes to Rosebery for a day's work ...

July 24 Dad left for Rosebery at 5:15. Joe and I did the chores. This afternoon we went down to the ice-cream social and also to tennis tea. GooGoo and I won our lone set. Sandy came down from the fire tonight. He is a pretty tired kid. They have evidently been having a pretty tough time of it up there, not the worst feature being that so much of their hard work seems hopeless and futile. A fire like that must cost the gov. no end of money.

July 25 I did some more teaming this morning. Tonight, Mrs. Hebron, Gertrude, Aileen, Heather, Euart Erickson and I went up in the car to Sandon to the tennis dance there. It was the first time I had ever driven a car over the road. I took exactly 1 hour which is just about the right speed I should say. It was quite enjoyable to drive over a totally different road for a change; one gets so tired of the 4 miles between Denver and Silverton. The dance was good fun. Mrs. Hebron our official chaperone was half the party. She certainly is awfully good fun. We arrived home about 5:30, had something to eat, and did not get to bed till 6:20 after Dad had got up.

The new road from New Denver to Three Forks, probably mid-1920s. Before this road, the only local drive by car was the four miles between New Denver and Silverton. Harris family photo

REFLECTIONS ON THE RANCH IN 1924

BY 1924 MY GRANDFATHER HAD SPENT MUCH OF HIS ENGLISH INHERITANCE. It had built a large farmhouse and paid the wages of the hired hands who worked on the property in the early days. The ranch so created was primarily an orchard, but also included a large kitchen garden and a variety of livestock: horses, dairy cows, pigs, chickens, ducks, geese, even trout in Loch Colin. There was a good-sized barn, a stable and carriage shed, a poultry house, and pens. Near the house were lawns and flower beds. This physical infrastructure remained in 1924, but with far less English money to sustain it. Nor was there much farm income. Judging by my father's diary, the farm sold a little butter every week and potatoes in late summer. Some eggs, fruit, and meat must also have been sold, but there is no doubt that the diary was written from a failing commercial farm. In these circumstances my father, my uncles, and my grandfather took on a good deal of non-farm work. Dad, Sandy, and sometimes Tod cut ranch timber, selling it to the Bosun Mine or hauling it to the lake (where it was boomed and pulled to a sawmill). Sandy also worked on the road from Silverton to Slocan, and Dad as a teamster in New Denver. Even my grandfather hired himself out with his team. They were working for wages when, as Dad said, there was not enough paying work on the ranch. The weekly films at the Bosun Hall, which my grandfather owned, and the public dances there, grossed little enough: $26.50 after one film, nothing after a dance because my father could not bring himself to collect. There was a new car, a Chevrolet, yet Dad worried about the cost of gasoline for runs from New Denver to Silverton (less than four miles). Money was very tight. The ranch had become what economic historians would identify as a struggling economy of multiple occupations.[1]

My grandfather's vision of a large, successful commercial orchard along a narrow bench of land above Slocan Lake had failed. An immense amount of work had gone for next to naught. Some of the fruit trees were diseased, but the orchard's failure stemmed principally from problems of marketing and planning, as noted in chapter 3. The ranch orchard had been planted when many of the Slocan mines were active and seemed to offer a sizeable local market. My grandfather had intended to supply it, and to this end planted many varieties of early and late ripening apples, as well as plums, pears, and cherries. From cherries in early July to apples in the late fall, he thought to supply the mines with fresh fruit throughout a long season. However, most of the mines closed, other, smaller Slocan orchards were planted, and by the time the ranch trees matured there was no appreciable local market for fruit. The only option was to ship fruit away, but without local packing houses or fruit brokers, the Slocan was not competitive with the Okanagan, where both were at hand and fruit ripened two weeks earlier.[2] Moreover, with no road away, exports depended on freight rates by steamboat and rail, and the CPR offered favourable rates only by the boxcar. With its fruit ripening at different times through a long season, the ranch could not fill a boxcar at any one time. It grew fruit for a market that did not exist. My father told of selling plums for 15 cents a crate when the crates themselves cost 20 cents.

Slocan Lake from Harris Ranch, n.d. but probably late 1920s. This is the ranch's Far Field, on which my father picked stones. Silvery Slocan Museum, New Denver

In April 1924, my father was pulling out recently matured fruit trees, a "terrible thing to be doing," he wrote, given all the "effort and hope" that had gone into them, yet it was "almost necessary." Some of the trees were diseased "and they are nearly all trees which will pay back very little." The problem of work to no end, which the orchard exemplified, was on his mind when, asked by his father to pick up stones on a ranch field, he reflected on "a wasted, futile day," and the "uselessness" of it all. Work to prepare a field for a hay crop, but hay for what? For dairy cows? What would dairying bring in? The ranch produced work, not product. "It seems to me that here, we are not concerned nearly enough about our actual products. We are not particular enough about the milk after it gets to the Dairy and of what becomes of it all. If we are not producing stuff and selling it in pretty good quantities what is the use of all the work. From now on we have got to have an eye more to the returns before we begin any job." Even tucked away in his diary, these are strong words from my father, especially given their implicit criticism of his father. Yet in 1924 the ranch was becoming a dairy, and would remain so for most of the next two decades.

My father's diary contains almost nothing about domestic life, almost nothing about his mother, and not very much more about his father. The household he had returned to was too familiar for comment. Far more in focus are his various connections with New Denver and Silverton.

Dad sang in the Methodist choir, and served on church committees. There were two team sports: hockey in winter, weather permitting, and baseball before the end of March. Dad pitched for the New Denver team, Sandy caught. Throughout the summer there was tennis—and tennis parties and tennis tournaments—on courts in New Denver and Silverton. There were frequent dances from which Dad would return in the small hours and berate himself next day for lack of sleep. There were political meetings to attend during a provincial election. There were many rehearsals, mostly wretched according to my father, for the play he directed. And there were those who came for visits, and often for supper—most frequently Mollie and Gertrude, both of whom usually stayed on for an evening of music.

Demanding as it was, the work of the ranch was situated in this matrix of local activities. Although the ranch was not yet thirty years old, it was not, like most pioneer settlements in Canada, a product of a stark encounter between poor immigrants and forest along a northern margin of North American agriculture. At hand, certainly, was forest and a pressing agricultural boundary, but also an English upper-middle-class background, and associated money, education, and expectations. Dad and his brothers and sister were born on the ranch and went to school in New Denver, but when the high school teacher proved an ineffective drunkard, my grandfather hired a teacher, Nancy Hill, a Harris cousin and recent graduate in classics from Cambridge, and established a school on the ranch. Dad, Sandy, Gertrude, and several other local children attended it. There, on a flank of a Kootenay mountain, Nancy Hill taught my father Latin, and convinced my grandparents and the Smiths in New Denver to send him and Gertrude to university. When Dad arrived at McGill, he knew far more Latin than any other first year student.

In 1924 the English money was largely gone, but not some of the ways that accompanied an upper-middle-class, English Methodist family

Dad, geese, and the new Chev in front of the ranch house, probably 1924. Probably taken on Sunday, when there was no ranch work, the photo seems to suggest that my father was ready for a drive to New Denver and a game of tennis. Harris family photo

inclined to social and spiritual reform. And so it was that my father cut wood, hauled stones, and shovelled manure by day, and by night wrote on the prefaces and plays of George Bernard Shaw.

It was no accident that Dad wrote a thesis on Shaw, whom both he and his father much admired. McGill gave him no help, not even an adviser. Left to his own devices, he identified the main topics of Shaw's thought and, after an introduction about the state of drama in Shaw's time, devoted a chapter to each: romantic idealism, medicine, war, sex and the family, economic and social theory, and religion. He even wrote to Shaw, seeking some clarification, and received the discouraging reply that one who would interpret an author's ideas more clearly than the author himself had no need of the author's advice. Dad was hurt, but thought Shaw was probably right, and never held his thesis in much regard. Yet he was steeped in, and had thought much about, all of Shaw's many plays and prefaces, and most of his other writing. And he wrote well. His thesis is a clear, thoughtful summary of Shaw's main ideas. I have the longhand manuscript before me now, the copy he sent to his typist. Alongside the ranch work, the play and its recalcitrant actors, too many late nights at dances, and an excruciatingly painful tooth, the thesis had been a huge struggle. But he finished it, sending it off at the beginning of April. If it is read alongside his diary, something of the soul of the ranch comes into focus.

Fundamental to Shaw was the conviction that the society he lived in was broken. It was often corrupt and always unjust. It produced and maintained poverty, the principal source of human suffering. Having no faith in the market's ability to produce justice, Shaw held that democratic government, beholden to the people, had to create the rules by which people could live equitably with each other. The poor had too little money, the rich far too much. The solution was equality of income; then the need for money would dissolve, competition would diminish, and people would be free to live decent, comfortable, and independent lives. "Drifters" hardly warranted such support. The wealth of a country should be divided among those who provided by their own exertions the equivalent or more of the goods and services they take. People should seek to leave society in their debt when they die. The source of such zeal was religious. The life force inherent in nature inclines to the good, but individuals can choose to attach themselves to it or not. In taking up, in willing, the cause of social justice, people attached themselves to the life force, to the essential teaching of Jesus, to the Kingdom of God on earth.

These ideas resonated deeply on the Bosun Ranch, and some of them crop up in my father's diary. Shaw himself was only partially their source. They were part of the English Fabian world in which both Shaw and, to some extent, the ranch were situated. In their light, the ranch was less a failed commercial farm than a locus of thought and argument about social change. A new and beautiful place seemed, at least to my grandfather, an opportunity to shape a better society. In his many pamphlets, letters to the editor of the *Nelson Daily News*, and radical songs sent to Paul Robeson, and year after year in his Sunday school teaching, he gave much of his thought and energy to this end. My father had grown up within this social project.

One of Dad's chapters considered Shaw's views on sex and marriage. Shaw thought that, for all the male conceits to the contrary, women were the hunters, men the pursued. In many of Shaw's plays, my father wrote, a "vigorous youthful female [was] busily engaged in tracking down her prey." Women hunted partly because of their vulnerable social and economic circumstances, but also because the universal life force—the "blind fury of creation" as Shaw put it—was particularly concentrated in them. "Woman," Shaw wrote in *Man and Superman*, "is Nature's contrivance for perpetuating its highest achievement. Sexually, man is woman's contrivance for fulfilling Nature's behest in the most convenient way."

What my father, still only twenty-three and probably sexually innocent, made of these views, and what he made of the women in his life, are questions without clear answers. GooGoo, whom he took to some dances and on short drives in the new Chev, was a pleasant if gossipy companion, but Dad was not interested in her. The "hot babies" he encountered at a dance in Sandon were not his type either. After his years at McGill, he found that the women he got along with in New Denver were educated and married. They were friends. But Mollie, Gertrude, and Kath Evans, with whom he corresponded? Where did they fit in?

Dad described Mollie as "a little ray of sunshine, always full of fun and good spirits," but he never took Mollie out, and Mollie and Sandy became a pair. They married in 1928, and moved into the cottage on the ranch, enlarged by this time, that Charlie Raeper had built in 1902. Their daughter, Nancy, was born in 1930. Gertrude was as frequently at the ranch as Mollie, and Dad was a handsome, educated young man (a rare commodity in New Denver in 1924)—quite possibly she had a romantic interest in him. If so, it was not reciprocated. Dad would have supper at the Smiths' then come home to write to Kath. He had met her at McGill, and they corresponded for years; clearly she fascinated him. A good letter from Kath left him "tickled and bucked up and stimulated and everything else." He said that she had "brain" and could write a "corking" letter.

Ellen Code at the time of her marriage to my father. Kenora, Ontario, 1930. Harris family photo

He was right. Many of her letters to Dad survive—not, unfortunately, those from 1924—and, full of wit, opinion, literature, and affection for Dad, are engrossing reading even now. She came from a prosperous, Conservative Winnipeg family, and lived with beaux, good conversation, and parties. When she moved to a job in Ottawa her crowd included Maryon and Lester Pearson. She came once to Vancouver to visit Dad, and he took her to the ranch. She told Dad that he was the only man she had never said no to, and one has the impression that she would have married him had he asked. But he never did. They corresponded for years, often about the state of their "friendship" and when and where they might get together. When he seemed out of sorts, she suggested he find a little local romance. In her last surviving letter to Dad, she is being pursued by a handsome and eminently eligible man, a Rhodes Scholar, something of a musician. She is not very interested but will give him a chance.

AS MY GRANDPARENTS AGED AND DIED, SANDY AND MOLLIE SETTLED INTO THE ranch and became its principal Harris family. Gertrude completed a Ph.D. at Berkeley, taught zoology at UBC, and in 1937 married Douglas Watney, a professor in the Anglican Theological College. Muriel, their daughter, is my wife. Dad took a teaching job in John Oliver High School in Vancouver. In the spring of 1929, Kath wrote to him about a Winnipeg acquaintance, Ellen Code, who was coming to stay with an elderly family friend in North Vancouver. Ellen, Kath reported, was very bright, very naive, and without young friends in Vancouver. She thought Dad would like her, and asked him to look her up. He did, and his correspondence with Kath ended.

At the Code family cottage on the Lake of the Woods in the summer of 1930, my father asked Ellen to marry him the following summer. No, she said, she would not, but she would marry him next week. Dad telegraphed the ranch to say that he was getting married, and a few days after the wedding, and a honeymoon terminated by Dad's horrified response to the cost of their room in the CPR hotel at Lake Louise, arrived at the ranch with a bride no one there knew anything about.

6 J. C. HARRIS, SOCIALISM, AND THE FABIAN IDEAL

AS THE GLOBAL ECONOMIC DEPRESSION DEEPENED AND METAL PRICES PLUMmeted early in 1930, the Slocan economy contracted sharply and there was virtually no chance of paid work for the unemployed. My father wrote to Ellen:

> MY LETTER FROM home this week tells me that mining is in a bad way up there. Nearly all the chief properties have closed down. The Ruth-Hope is about the last to close. That means that my kid brother Tod is out of a job, temporarily at least, and that Bud and Heather [Dad's sister] may have to leave Sandon as Bud was Superintendent at the Ruth-Hope. This mining game certainly is an uncertain one. The price of silver is lower than it has been for decades; lead and zinc are also down. Things certainly look bad for the Slocan now and a year ago they were enjoying what was almost a boom. In fact it looks as if there was over production in everything ...[1]

In these circumstances, the Bosun Ranch became an oasis. It was welcoming, and provided good food, accommodation, and plenty to do. My uncle Tod soon moved back, later my aunt Heather and her family. Sandy and Mollie already lived there. Men who came to work stayed as long as possible. One, whose job was to build a carriage shed, took so long that the shed became "the cathedral" (after the great European cathedrals that took centuries to build). Of the Harris children, only my father lived away. He had a teaching job in Vancouver and, although his salary was cut and missed entirely one month, it held him there—except in summers, when he and my mother usually did get to the ranch (chapter 7). Money was tight, but life there was good. Years later, my father remembered the Depression years on the ranch particularly fondly: "My wife and I both often almost envied the rest of the family and the good times they were having."[2]

By 1930, most of the orchard had been pulled out and replaced by meadow and pasture. The ranch had become primarily a dairy. There were usually six milk cows in the barn; the fields produced hay and silage.[3] Twice a week my grandfather ran a milk wagon to New Denver and Silverton. Family lore has it that those who could not pay got milk anyway. With only a few cows, Depression prices for milk, and probably some quiet philanthropy, the ranch cannot have produced much income, but my grandfather thought that the combination of meadow and livestock on his and other small Kootenay farms finally had some prospect of producing a viable economy:

> I DO BELIEVE that a very considerable livestock and dairy industry could be built up in these pleasant and healthy Kootenay valleys. It will always be a hard work country, steep side hills, rocky, and naturally poor soil that must be carefully built up even to get the successful first "catch" of alfalfa. Most of the farms and fields will naturally be small and irregular with bedrock outcropping and little creeks that can become raging torrents if not most carefully handled. Yes a country for real men who have the right cooperative spirit and who will love our mountain lakes and valleys even if a very large part is barren mountain, and magnificent wilderness.[4]

His words describe his own small, transformed farm, at which he continued to work exceedingly hard although, as he was now over sixty, with declining energy. He could be very absentminded: one evening, he took a lantern and bucket to the barn, placed the lantern under a cow, and began to milk. Often in these years, his thoughts were with broad plans for social change. His production of letters to the editor, booklets, pamphlets, and radical songs increased, and he lectured to whatever audiences he could find. Farming was not his preoccupation. My father feared that his father was becoming a fanatic; my mother thought that, living on the ranch, her father-in-law was far too isolated.[5] Certainly, as the world's economic order seemed to crumble, the times were drastic and radical options, including my grandfather's, were much in the air.

He had come out of a family of Gladstonian Liberals interested in social reform, and had been considerably radicalized while farming on Vancouver Island. He had read many of the English Fabian socialists, knew something of Marx, greatly admired the American economist Henry George, and was intrigued by technocracy's vision of a society run by experts. He was also deeply religious, and, as my mother intimated, thought about religion and politics on an isolated farm above Slocan Lake.

In the provincial election of 1909, he ran as an independent candidate. Hoping to acquaint electors with his views, he published a booklet, *British Columbian Problems*,[6] approximately quoting Francis Bacon on its cover: "Read not to contradict, nor to believe, but to weigh and consider." He began by considering mining. Following American economist Henry George, he outlined his thoughts on taxation, arguing that taxes on mining claims (land) should be raised to discourage owners from sitting on their holdings, and those on working mines reduced to encourage their development. Owners of claims should be required to establish a value for their properties (otherwise a government assessor would appraise them), and their properties would then be taxed on the basis of this value. Moreover, properties would be automatically sold to anyone willing to pay the established price plus 5 percent. By these means undeveloped mining claims, which he took to be the principal impediment to the development of the Slocan, would be fairly priced and quickly worked.

More generally, he thought that capitalism produced inequality and poverty, and that in the interest of social justice, government should centrally manage the economy. Forests, for example, "should be handled on the principles which govern a great estate. They should be developed and exploited as to yield a great and increasing return in timber, and not handed over to a few men to destroy. We should adopt the policy of a good farmer as opposed to the methods of a pirate or a highwayman." If owned by the Crown and scientifically managed, the forests would "reap the advantages of a well worked monopoly." Similarly, government should control and limit the distribution of population: "We have adopted an absurd land system and have allowed speculators everywhere to tie up the country. We have built our roads past lands that are lying idle, and our railroads are pushed all over the Province long before they are really needed. A wise Government would have insisted that before they opened up fresh sections of the country, the parts already occupied should be adequately peopled and developed." Unneeded railways through scattered populations were expensive luxuries that carried little traffic, charged excessively, and caused forest fires. "Canada is now like a big ocean liner that has started on a voyage with the crew of a fishing smack. This incessant craze for opening up new countries and hoping thereby to find prosperity ... is like the action of a small boy chasing the tail of the rainbow for the mythical bag of gold."

From belief in a managed economy followed the conviction that the quality of government depended on the quality of the civil service. Were the civil service "rotten and inefficient," government management would not work. Civil servants of outstanding ability and unquestioned probity were essential, and should be protected from the "unclean hands" of party politicians. Moreover, a socialistic spirit need be in the air, "that 'fellowship' which William Morris spoke of—that 'spirit' which was infused in Christ's life and teaching." This was critical. "To my mind the extent to which the principles of Socialism can be successfully carried in any country will for all time depend on the public spirit, honesty, and gentlemanliness of its citizens, and, even if the substance of Government ownership were carried by popular vote, without the spirit which should animate it, we should have made no advance in human happiness."[7]

Although he identified himself a socialist, he thought the small Socialist Party of Canada erred in excessive criticism of politicians from established parties (most of them were decent people), and of government (goodwill toward government was essential to socialism). He thought socialists effectively described "the horrors of the present system" and the glories of the socialist state, but provided "a most inadequate ladder ... by which we might advance to it, so that the average citizen feels as if he might as well take a tramcar to the moon." Often the socialist state seemed "about as real, human, and possible as fairyland." Nor did he think the entire economy should be nationalized.

In a short concluding chapter devoted to orchardists, he lamented the gulf between the price growers received for their fruit and what consumers paid for it, acknowledged that transportation costs were not the principal problem, and railed against middlemen: "The curse of the age is the immense number of parasitic individuals that have succeeded in foisting themselves on the backs of the real producers of wealth. Our miners, fishermen, farmers, and manufacturers have to carry this vast crowd of unnecessary agents."

Very few in the Slocan in 1909 can have bothered to "weigh and consider" these ideas. They did not speak to radicalized miners, horrified mine and claim owners, and were completely out of tune with the speculative energy that had developed the modern Slocan. My grandfather, however, was undaunted. Over the years, his basic views hardly changed, though they became much more generalized. By the 1930s, he had become a gentle, somewhat eccentric, and completely committed crusader for his broad vision of a more cooperative, efficient, and just society.

Behind this vision lay his deeply held Christianity. He had turned away from religion only briefly on Vancouver Island, there influenced by a well-read German neighbour (chapter 1). He was well aware of the arguments over Darwin, and having read something of the English biologist and philosopher Herbert Spencer and more of Marx, the beliefs with which he had grown up had been shaken. The book that seems to have drawn him back was *Natural Law in the Spiritual World*, written by the Scottish scientist Henry Drummond.[8] When he first read it is not clear, but late in his life he said that Drummond's arguments that the laws being discovered by science were also, and in the first instance, God's laws, restored his faith.[9] For Drummond, truth in Nature came from God. God had simply transposed his laws from the unseen to the material world. Nature, therefore, was the working model of the spiritual. Moreover, science was increasingly demonstrating that the laws of life and the laws of society were intertwined to the point that the whole material world was able to reveal God's handiwork.[10]

These ideas greatly comforted my grandfather. The Bible need not be taken literally, and, rather than undermining religion, the many achievements of science revealed God's design for the world. God remained in charge, if expressed now through natural laws. To be sure, He had become more remote. In a world governed by natural laws emanating from God and expressing His truth, what, for example, was the role of prayer? God's laws—nature's laws—worked as they worked. Drummond had opened up an ordered, almost mechanical realm not far removed from that envisaged by seventeenth- and eighteenth-century deists who held that God created and ordered the world, then left it alone. But in a late Victorian world hungry to rescue belief, Drummond's ideas had a great deal of purchase, as they did with my grandfather.

He was also steeped in the Bible, and deeply committed to the moral teachings of Jesus. We are "our brother's keepers," and we "bear each other's burdens." Loving, giving, and sharing. Forgiveness before judgment. Kindliness and consideration. Concern for the poor. So the "carpenter of Nazareth" taught, so we should live. However tamed by the rich, Christianity, he thought, was essentially "a religion of revolt ... founded by the noblest of all revolutionaries."[11] And, in his eyes, a social revolution was desperately needed. He had left a corrupt and decadent society, had come to a beautiful valley only a few years after its modern beginning, and saw there the glistening gift of God and the opportunity to create a far more just and caring society than he had left behind.

He created a political party, the Useful People's Party, ran under its banner in a provincial election, and apparently received three votes. With neither organization nor formal membership, the Useful People's Party was essentially a vehicle for the expression of his convictions that society was far too committed to the idea of liberty at the expense of duty and responsibility, and was divided between useful and useless people. For their own and the social good, the latter, he thought, should be put to useful work. These views appear over and over in his unpublished political essays, his letters to the editor, and his Useful People's Party publications. Years after he joined the Co-operative Commonwealth Federation (CCF), which he did at its inception in 1932–33, he encouraged his readers to study the Useful People's Party publications and bring their ideas to whatever political party they belonged.

At the same time, he was soon deeply involved with the CCF, attended at least one of its conventions, and knew most of its leaders. J. S. Woodsworth (the party's first leader), M. J. Coldwell (its second leader), Lyle Telford (a founder of the party in BC and briefly mayor of Vancouver), Grace MacInnis (founder) and Angus MacInnis (founder and member of parliament) all stayed at the Bosun Ranch when campaigning in the Slocan. Although Joe readily embraced the CCF and its vision of a "cooperative commonwealth," he was also a critic. He thought that arguments within the party about the nature of socialism and the means of achieving it blurred its message and weakened its prospects. As in 1909, he thought socialists far better at identifying the failings of capitalism than in providing a clear and accessible alternative. And he considered many in the party too cautious, too caught up with what Sidney Webb, a founder of English Fabianism, considered "the inevitability of gradualness." Drastic times called for drastic action, and my grandfather had a plan of sorts that he thought the CCF, and Canadian society more generally, should adopt.

Those who attended his addresses on "Cabbage Politics" in the "Vacant Lot Opposite Sargeant's Garage," in New Denver on May 28 and 29, 1937, heard parts of this plan, and, as a tattered typescript survives,

Useful People's Party

Are You A Really Useful Person?

Why Is It So Hard To Be a Really Useful Person in Canada Today?

J. C. Harris of New Denver

Will Address Public Meetings On

Wed. and Thur. Evenings

MAY 28th and 29th

In the Vacant Lot Opposite Sargeant's Garage, Commencing At 7:30 p.m.

His Addresses Will Be On CABBAGE POLITICS

You will find "Cabbage Politics" kindly, humoursome, and easy to take on a moderately full stomach.

All disagreeable after effects can be removed by talking them over with your neighbours and friends.

CAUTION

The scientific method of attending a Useful People's Party meeting is to put a nickle or even a dime in your pocket and hold on to it tightly. If the speaker does not fairly earn your money by talking good sense and giving you a good time, return your money to your safe.

Ladies especially invited, in fact the speaker rather prefers the ladies.

Handbill for the "Cabbage Politics" lecture, 1937. It takes a particular optimism to print a handbill for a lecture to be delivered in a vacant lot in a Kootenay village of some three hundred people. Harris family papers

J. C. Harris with the hand cultivator he used to battle weeds, mid-1930s. After some forty years in BC, my grandfather still wore a tie when farming. Harris family photo

there is no doubt about what he said. He offered a transformative social vision. Times, he said, were exceedingly difficult, problems immense, and commonly suggested remedies trivial. People "not used to worrying about their daily bread find themselves forced to make painful retrenchments and humiliating economies." In these circumstances, he turned to nature. She taught stern lessons that farmers learned and city folk, caught in the bustle of the city, could hardly hear. His was plain speaking from "an old hayseed," and it had become his habit to compare humans to cabbages. Just as there were many varieties of cabbages, so there were many varieties of humans. Whatever the variety, the value of cabbages, as of humans, depended on their having good sound heads and tender hearts, and the challenge in any society was to produce as many people with sound heads and tender hearts as possible. Yet in Canada there was a "dearth of noble natures."

Having spent much of his life weeding, he suspected that weeds—"the unruly mob that infests our gardens, or the misplaced individuals that infest human society"—were the problem. That spring he had absent-mindedly set out cabbages on ground already planted in peas. Coming up, they were weeds, the one to the other. Having realized that, out of place, the noble cabbage became a weed, we "can understand that human beings, sown broadcast as it were, in society today, are bound in very many cases to be weeds also." Individuals were not at fault. He himself had "laboured hard to produce good fruit and had shipped my product in high hope, only to find the market glutted and that I, to the extent that I had laboured to produce unwanted food, was a miserable weed." Such an experience made him tolerant, but did not lessen his indignation at society's muddle and confusion, present ideals, and "barbarous estimate of success in life." Valuing freedom and liberty, society assumes that its interests are best served by allowing the individual to do what he or she likes best, but this, my grandfather held, is to assume that the fittest will survive. It would not work with cabbages, which have to be protected, watered, and cultivated, nor will it with humans:

> IS IT LIKELY that we shall get good sound heads and tender hearts by the methods of the jungle? Cannot we arrange that each human cabbage should be properly placed and spaced and even transplanted to more favourable conditions if it seemed necessary? ... To a farmer who has once grasped the idea that cabbages and human beings are of marvelous similitude, the idea of transplanting and setting out the human cabbages seems the most natural thing to do, and our present method of muddling along seems ridiculously like sowing cabbages broadcast and allowing them to take their chance.

No other young people, my grandfather thought, had been treated as generously as those of his day in North America. However, nothing was required of them other than to enter the competitive struggle. No one had said, "Unto whom much is given, of them much should be required." Even should this teaching be heeded, the young could hardly know how or where to find useful work. They needed advice and direction, and society, my grandfather thought, had "the right and the duty to say to each and all of the rising generation: you must try to do your full share of the work that your country requires to be done." It would try to allot work that was as agreeable and interesting as possible, but there were jobs to be done, and just as in wartime people were conscripted to do them, so in peacetime they should be conscripted to work for the common good. Useless work, the bane of modern life, would end. Those who refused to be conscripted would be denied the use of public facilities—the roads, for example, except to walk on.

North America, my grandfather thought, resembled a huge engine of about fifty million manpower without a governor. "It still trusts ... to its once sufficient governors, supply and demand, but these old fashioned contrivances are very much discredited by the results obtained." A far more collective vision was at hand: "Fancy feeling yourself part of a great organism that was working intelligently and systematically to supply all mankind's reasonable needs. Fancy the great burden of apprehension being lifted from trembling mankind as we gained courage from the feeling of human brotherhood and mutual support." In such a society, priorities would change. "It is a national repentance that I am pleading for. A recognition of the danger that we are in from our admitted success in the production of commodities; a realization that the production of things is not of much importance in comparison with advance in the supreme art of living."

Earlier that year, he put the same ideas in a short booklet, *The Cook's Strike*, a Useful People's Party publication. It contained the correspondence of two young Ontario women, Alice and May, with their elderly, bedridden, suffragette aunt Dorcas in England. At a meeting of the local Ladies Aid, the chair, Mrs. Stately, had read a letter from an overwhelmed farm family in Saskatchewan to whom the Ladies Aid had sent used clothing the previous year. Some of the family, it turned out, were now dead and others were dying. "We ladies," reported Alice and May, were "wiping our eyes and trying hard to look as if we were only blowing our noses, when a most penetrating and arresting voice broke in sharply, from the very back of the room. 'Well, folks. What are you going to do about it.'" This was Mrs. Knap (Gingersnap as she became known), who was soon on her feet again to say that there was enough food and clothing in Canada to make every family comfortable. She had grown up on a farm. She knew that poorly fed cows gave little milk. Women in Canada should take their cue from the cows and go on strike—the Cook's Strike, or the Great Apple Pie Strike as the press later called it—thereby supporting needy families and building pressure for change. The press was horrified—"RIOTING WOMEN WRECK DRAWING ROOM"—but the movement was rapidly gaining momentum. Gingersnap was its leader, and Alice and May were in the thick of it.

Her nieces' letter, Aunt Dorcas replied, had been a wonderful tonic. She had bounded out of bed (she was still suffering the effects of a police charge of suffragettes at No. 10 Downing Street), and wanted her nieces to give Gingersnap kisses from "an old warhorse who scents the battle from afar." She then offered advice that sounded suspiciously like my grandfather: get rid of uselessness, adopt conscription and organize society to be useful, deny access to public services to those who will not work for the common good, and understand that duty overrides liberty. The nieces, for their part, were delighted to hear from their aunt. The movement, which had started with so much enthusiasm and energy, had sagged. The strike had been postponed, although many Canadian women were "deliberately cutting down on their men's dinner (leaving out the pie)." Now Aunt Dorcas had reenergized them. They had adopted her ideas about conscription, and were mobilizing the entire nation to fight "our real foes—uselessness, idleness, greed, ignorance, and low ideals."

In these and many other writings, my grandfather held that society should be organized and managed, and that work should be assigned to serve the common good. In so doing, he thought, useful work would replace uselessness. But what is useful and what is useless work? Who, or what organization, is to decide, and by what criteria and by what right? These are basic questions, and the closest he came to answering them was in his booklet *Conscription for Peace*, the first edition of which was also published in the late 1930s.

This booklet reiterates the arguments for peacetime conscription that he made in his speech "Cabbage Politics"; suggests that Canada should be organized as a well-run mixed farm; and holds that Canadians should commit themselves to cooperation, brotherhood, and common

Cover of *The Cook's Strike*, a Useful People's Party publication, 1937. The cover of this publication conveyed its essential message: the excessive worship of freedom entailed competitive, destructive human relations for which the cure was a strong sense of social responsibility that would lead to a compassionate and cooperative society. Harris family papers and J. C. Harris fonds, Royal British Columbia Museum and Archives purpose in a centrally managed economy. But how to accomplish this momentous transformation? My grandfather, who was a critic of socialist fairylands to which there seemed no access, may have created one of his own. Yet, he did have the edge of an agenda: within a hundred days of its election, a supportive government would pass a law requiring all able-bodied men and women between the ages of twenty and fifty to join a trade union or farmers' institute. Bankers and some other professions would probably have to create their own new unions, as would merchants. People could join whatever union or farmers' institute they wished but, having signed on, would agree to undertake the work assigned to them. Committees of experienced workers in each union or institute would decide on the work to be done and, as far as possible, fit individual abilities and preferences to particular tasks. No force would be used. Those who refused to sign, or refused to undertake assigned work, would simply be denied the use of public facilities. So shamed and inconvenienced, they would probably soon come round, and when they did they would be fully reinstated.

The effects of this reorganization would be many and salutary: the ownership of the means of production would shift to the workers; class differences would virtually disappear; people would begin to pull together, part of one large organism and united in a common cause; many more, working through their unions and farmers' institutes, would participate in the art of government; and all would be paid for what they did rather than for what they owned. As a class, rich folk and former bosses would quickly disappear; meanwhile they would be treated with kindness and patience. In so doing, our "great, rough, lovely Canada" would be put in order.[12]

He was never more specific. His goal—a reorganized society based on cooperation and brotherhood—was clear, but his agenda for social change was never much thought out, and his few plans seem implausible and unworkable. Yet he was far less interested in means than ends, and the end he offered was a moral vision, one probably very close to his sense of the Kingdom of God on earth.

He wanted criticism, and received it, particularly in letters to the *Nelson Daily News*. Both he and his critics had found an obliging editor who apparently published all their letters,[13] however lengthy. Most of the critics did not give their real name. "Citizen" wrote that conscription during wartime had not been a model of efficiency. The logistics of supply and troop movement had often been mishandled. Discipline among troops was always difficult to maintain; indeed, the last war had been won because the Germans mutinied first. "State socialism that for its success requires the obedience of every man, woman, and child, would more speedily develop open mutiny and defiance. It would end speedily in disorganization and chaos." "Canadian" wrote that while he appreciated Mr. Harris's many references to Jesus of Nazareth, Harris had completely misinterpreted Christ. Jesus had limited his teachings to spiritual matters. His primary mission on earth was to suffer, and thereby redeem the souls of men. "Moderation" considered it idle to look for churches, public services, and care of the poor in a Christian community that fails to exercise ordinary business prudence. J. T. Bealby wrote that he agreed with Mr. Harris that we must seek the Kingdom of Heaven and renounce our worship of Mammon; indeed in several letters to the *Nelson Daily News* the previous September he had said as much himself. But Mr. Harris had been too hard on capital, he felt, and had drastically underestimated the complexity of the country he purported to plan. Human nature, far more self-serving than socialists admit, had always been the bane of socialist schemes. Eventually, my grandfather and J. T. Bealby argued over whether Jesus encouraged accumulation. My grandfather was certain he did not. Bealby said he did not need to because Palestine was a land of milk and honey. In several long letters to an accommodating newspaper, the argument then turned to the frequency of famine in biblical Palestine.

These were arguments from the early 1930s. A dozen years later my grandfather wrote a series of letters to the *Nelson Daily News* in which, imagining himself an astral spirit, he revisited the town of Trail after an absence of ten years. It was very much changed, for which the impetus, he learned, was a horrendous, worldwide economic crash in the year

1947. Governments had called out their armies and virtually put them in charge. Rioting was checked and necessities rationed. Certificates for work or relief became the currency. Economies were managed for the common good. Cooperation replaced competition. The system of rents and profits, "the great stronghold of uselessness," was overthrown, and with it the rights of private property. He met with John Bold, the mayor, who reported a new spirit in the air. "We have put glory into drudgery, we have put pleasure into the common business of life, and poetry and the enthusiasm of purposeful creation into our meanest task." Workers, he said, now managed the smelter, the bonds between management and workers were therefore close, and, as everyone had enough, there was little difference in pay.

A transformed economy was apparently booming. A new canal and locks ran through Trail, turning it into a modern port. Tourism was flourishing. The smelter was far less polluting. The town looked better: more parks, more community centres and playing fields, more fine municipal buildings, more public baths and wash houses. Inner-city Trail had shrunk, and the suburbs had grown, the two connected by trams and shorter hours of work. Well-kept lawns and bright flower beds. People at ease and happy. My grandfather offered a utopian dream, much as William Morris dreamed in *News from Nowhere* of a remade and de-urbanized London where, free from the constraints of property rights, people lived beautiful lives in beautiful places surrounded by a beautiful nature.

ALTHOUGH IT HAD FEWER THAN A THOUSAND MEMBERS WHEN MY GRANDfather joined in 1896, the Fabian Society included many of the prominent British intellectuals of the day: George Bernard Shaw, Sidney and Beatrice Webb, William Morris, Havelock Ellis, for a time H. G. Wells, and many others. They wrote voluminously and argued about details of the socialist enterprise, but all of them held that socialism could only be achieved democratically, and that education rather than revolution was the means of achieving it. Capitalism, they assumed, had run its course. Its contradictions were obvious, its underlying greed amoral, and its production of poverty intolerable. A higher, more cooperative mode of social-economic organization was almost at hand. To achieve it, land should be nationalized; workers should own the means of production and receive thereby the surplus value of their own labour. Most Fabians envisaged scientifically planned societies run by skilled and benevolent experts. Political parties, as such, would probably disappear (Sidney Webb). As there was no logical basis for assessing relative worth, everyone—women and men, judges and carpenters—should be paid the same (Shaw). Rescued from the competitive struggle for money, people would be freed to enjoy their lives. A classless society would emerge. Women, no longer dependent on their husbands' earnings, would escape the patriarchal power inherent in traditional marriages.

Such a society would confer many gifts, and with them would come the responsibility, the duty, to pay them back, even to return more than had been received. To this end, Shaw advocated a form of compulsory national service. Underlying all such Fabian planning was the premise that socialism would replace greed with cooperation. Bourgeois culture was spiritually sterile (Morris). Moreover, capitalism was using machinery to dehumanize people. In freeing them to exercise their own creative talents, socialism would restore people's real worth and transform ugliness into beauty. Perhaps Morris's most basic belief, one he shared with virtually all the Fabians, was that a cooperative spirit was deep in people, and that, freed from the degrading scramble of capitalist society, it would uphold the sense of brotherhood and common purpose required to undergird a socialist society.

While still a fairly young man, my grandfather encountered and largely embraced these views and, with minor variations, held them for the rest of his life. Basically, his was an English Fabian voice emanating from a Kootenay mountainside. There, most of what he had to say sounded exceedingly strange, but would have been quite orthodox in English Fabian circles. Because he was a farmer, he used metaphors of cows and cabbages and well-managed mixed farms, but the message did not change. Perhaps his exposure to technocracy in the 1920s and

’30s, and to its claims that society should be run scientifically by engineers and technical experts, increased his enthusiasm for planning and management, but he was familiar with similar views in the Fabian literature long before he encountered technocracy. Probably his distinction between useful and useless people was not quite as English Fabians would have drawn it. For many of them, uselessness was a product of poverty, whereas for my grandfather it was associated with the crowd of largely urban speculators, sharpies, and middlemen, many of them much too well off, who lived off the avails of working people. He had lived almost all his Canadian life on a farm, held urban life in low regard, and his sense of uselessness contained a good deal of rural Canada’s deep distrust of accredited experts and city slickers—people with few practical skills, inflated egos, and strong senses of entitlement. Conscripting people, the useless among them, for peace was but a form of the compulsory national service advocated by some English Fabians.

In many ways, this Fabian legacy had been radically decontextualized, just as it had been for R. B. Kerr and his wife, the other Fabians in British Columbia in the mid-1890s, and the pair who encouraged my grandfather to settle in the Slocan. Kerr was a lawyer, and when twelve prostitutes were charged in Kaslo (a village on Kootenay Lake almost thirty miles from New Denver) he argued that in a predominantly male mining camp, prostitutes were essential for social stability. If any, the men who frequented them should be charged.[14] Most Fabians in England would have agreed with him, but Kerr was in a western North American mining camp where almost everyone, except perhaps the women in question, thought him daft.

Similarly, my grandfather sought to introduce a broad socio-economic agenda created in one place to another that was radically different. The modern Slocan had been built on speculative energy, on schemes and greed, and was a most unlikely receptacle for a body of English ideas based on the inherent goodness of people, their bent toward cooperation, and their willing participation in a centrally planned and managed economy. Beyond the Fabian circle, such ideas were much contested even in England; in the Slocan they had no constituency.

My grandfather with my sister and me, summer 1941. Beneath my grandfather's political rhetoric, and perhaps long outliving it, was his essential kindliness. Here he takes two of his grandchildren for a wagon ride. Harris family photo

Moreover, the Fabian ideas that captivated my grandfather were developed in the three decades between the founding of the society in 1884 and the beginning of World War I. Then, naivety was still possible—before the horrors of that war and of its successor; the brutalities that, in the name of the common good, governments, particularly in Russia and China, meted out on people they governed; the ease with which campaigns to improve public morality tended to become horrendous purges of difference; and the impenetrability and venality of many bureaucracies. Today it hardly is, and for that reason many of the reforms my grandfather proposed now seem unworkable, even dangerous. He often said that brotherhood and sharing each other's burdens were "the law of Jesus," and held that God had given his utopian community of Trail "a good push from behind." But if, after Henry Drummond, God had transposed his laws from the unseen to the material world, if divine laws ran through societies and could become particular political agendas, what did one do with dissent? One enters a realm of truth and error with the horrible consequences that could so easily follow therefrom.

This, however, is more easily said now than a century ago. My grandfather was a kindly man who strove to create a more just world based on cooperation and sharing. He gave an enormous amount to this undertaking: hundreds of letters to the *Nelson Daily News*, Sunday school classes for fifty years, his political pamphlets, the Useful People's Party, work for the CCF, his political songs, talks like those in the vacant lot opposite Sargeant's Garage. That is how I think of him, trying, with a fanatical edge softened by kindness and humour, to convince whoever would listen that greed should give way to cooperation and that we should all work for the common good in wisely managed societies. Looking over all this near the end of his life, and also at a world that had not budged for the better, he told my parents that he had failed.

7 THE LOG CABIN

FOR ME AS A BOY, A LOG CABIN LOCATED ABOVE LOCH COLIN WAS THE HEART of the ranch. Almost every summer after the war, my parents, my sister Susan, and I spent the better part of two months there. With a wood stove, an icebox, and coal oil lamps, the cabin was rustic but also elegant, and in my mind as beautiful as a building could be. My sister's and my Vancouver friends often visited, as did relatives and adult friends of my parents. Loch Colin was cold but inviting, and we played on and around it endlessly. There was always talk, always a new project underway. And there were rituals, among them at the end of each summer my father measuring our heights on an interior post. I loved the cabin, partly because it was such fun to be there, partly because it was so beautiful. Years later, Muriel and I spent our honeymoon there.

Moreover, the cabin accumulated stories, the most repeated, perhaps, about Mr. Weston and the skunk. He, an artist of some renown, was painting the early morning sunlight on the mountains across the lake without realizing that a skunk lurked under the cabin, caught in my gopher trap. When I discovered this striped gopher, and Mr. Weston identified it, and Aunt Mollie came and shot it, the skunk took its revenge. A wall of scent moved across the cabin. Mr. Weston retreated, painted furiously, and then, holding his nose, would dash out for another look. When the wall reached his retreat, he gave up. All of us moved out for a week. The painting is not one of Weston's best. In my young eyes the cabin had always existed, but in fact it had just been made. Planned in the late summer of 1932, it was begun in 1933 and built over the next eight summers. Dad did much of the work, but the idea, I am sure, was my mother's. She

Mother at Loch Colin, 1932. Pond family photo

wanted a place of their own, apart from the ranch house, yet close by. They were deeply in love, and were creating a love nest.

Mother and Dad were letter writers, and when apart wrote to each other almost every day. Most of these letters survive, and I hoped they would contain information about the building of the cabin. But they don't. The two of them were together on the ranch when most of the work was going on, and when they were not (because Mother was visiting her family in Winnipeg, or because she was ill there and Dad was in Vancouver), they wrote about other matters. Mother's album of cabin photos, an estimate of labour and material costs, and a few stories are what survive about the making of the cabin. They permit the following visual chronology of its construction, but say nothing about the aesthetics and values embedded in this remarkable building.

1932: Mother spent most of this summer in Kenora and Winnipeg, and she and Dad were together on the ranch only in late August, which is when they probably decided to build a log cabin just above Loch Colin on a tree-covered knoll. The site they chose (to Mother's right in the accompanying photo) was tucked away above and out of sight of the ranch house and about 150 yards from it.

My father, an unidentified man, and my grandfather (left to right) at a corner post, summer 1933. Pond family photo

1933: Dad cut, peeled, and hauled (using a horse) white pine logs from the hillside above Loch Colin to the building site. By the end of the summer there was a fair pile. Concrete corner posts like the one in the accompanying photo were poured and the first two rounds of logs laid.

1934: Dad got two more rounds of logs in place. There were only a couple of summer months to work with, so the cabin progressed slowly. Dad, who was good with an axe, notched and fitted the logs; Mother came to admire.

1935: Dad got the log walls above door height. By the end of a third summer's work, the cabin's main room had almost emerged. A door had been cut out, and spaces left for cutting out the windows. The logs were spiked together as well as cross-notched. Bud Rose, Dad's brother-in-law, split shakes, and in the fall a carpenter from Nakusp (not identified) framed and finished the roof and did a good deal of interior work.

Here is the carpenter's "Estimate for completion of log house for Mr. R. Harris, at the Bosun Ranch," dated August 30, 1935:

My father, mother, and dogs inside the emerging shell of the cabin, late summer 1934. Pond family photo

MATERIALS

- Shiplap for floors, roofs, etc. floor joists and plank for frames, 4,200 ft, del. approx $113.00
- Casement sash, del. approx. 46.20
- tarpaper 4.50
- galv. iron for valleys, shower stall etc, and solder. 12.00
- Nails, common, various sizes 9.00
- bricks and lime for kitchen chimney. 17.00
- casement hinges, door hinges, fastners, and general hardware, approx 18.00

materials $219.70

TO LABOUR AS DETAILED

- log and carpenter work $272.00
- plumbing complete, lab. only 45.00
- septic tank, lab. and mat 43.00
- finish floor
- coast fir, labour and materials 88.00
- local flooring, lab. and mat 59.00

Mother with Jenny (Scottish Terrier) and Wapps (English Springer Spaniel) at the cabin, late summer, 1935. Pond family photo

To labour for raising side walls of living room, raising four gable ends, fitting purlins, ridge poles, valleys and rafters, filling between rafters, building end wall of living room, building bathroom walls, this to be all log work, roof to have wide eaves and good overhang at gables

Openings to be cut out for casement windows, frames to be substantially built of plank and sashes well fitted and hung to open outward.

Door openings as above, outer doors, "Dutch" style, inner doors one piece, built up construction.

Floor joists to be leveled on stringers and first floor of shiplap laid diagonally.

Roof to be close sheeted with shiplap, covered with tarpaper, and strapped with 1″ × 4″ then covered with split shakes.

I estimate cost of labour for above work @ $272.00. I estimate cost of plumbing complete, viz. shower bath, toilet, basin, hot water tank, kitchen sink, all necessary pipe work, labour only @ $45.00. for septic tank, labour and materials @$43.00.

All materials to be delivered at the site of building and shakes to be split.

This estimate anticipates the completed cabin, and describes an enormous amount of skilled work—final log work, particularly in the gable ends; the whole structure and finish of the roof; walls around the bathroom; flooring; plumbing; the kitchen chimney. Much of it was accomplished in the fall of 1935. Apparently the carpenter did not then make the frames for windows and doors, or the doors themselves. In an accompanying note my father indicates that he had ordered the window sash separately.

Overall, it is apparent that by the summer of 1935 Mother and Dad had worked out the principal details of the cabin down to the hardware of windows and doors, and had found a remarkable craftsman to do the work. There was never more than a rudimentary plan, although the three of them must have talked about eaves, overhangs, and other details. I sense a meeting of minds; the carpenter and my parents—particularly my mother—seem to have shared a vision.

When Mother and Dad returned to the ranch in the summer of 1937, they must have been delighted by the interior spaces they found and the quality of the workmanship. There are no family stories about the carpenter from Nakusp, decisive as his work was, probably because when he worked on the cabin my parents were in Vancouver. While growing up, I never heard of him.

1936: I was born. Mother and Dad stayed in Vancouver.

1937: Mother and Dad returned in the summer of 1937 to a building that was almost habitable. It had a roof, floors, an enclosed bathroom, some plumbing, and a brick chimney in the kitchen. However, the full openings for the windows were not yet cut, there was no log partition at the interior end of the living room, and the fireplace was not yet built. The principal improvement in the summer of 1937 was the fireplace.

It was built by Mr. Pomquist, an elderly Swede who lived near Slocan City, and about whom the following tale is part of the lore of the cabin. During prohibition Pomquist lived well apart from the town in a

Mr. Pomquist's fireplace, 1937. Pond family photo

hillside cabin with his goats and a still. When the police got wind of the still, they sent two men to arrest him. Fearing that he was dangerous, one hid behind a tree while covering the cabin with a revolver, and the other knocked at the door. "Come in, come in" said Pomquist, "Come in and have a drink." The officer explained that that, in fact, was the problem. That drink and the still were illegal, and he would have to arrest Pomquist and take him to jail in Slocan City. "No," said Pomquist, "I can't go, I can't leave my goats." In the event, both Pomquist and his goats went to jail. There was a trial, Pomquist was guilty, and the judge fined him fifty dollars, which Pomquist could not pay. The town took up a collection and raised forty-five dollars. The judge put in the last five. Pomquist and his goats went home.

Almost two decades later, when Mr. Pomquist arrived at the ranch, he sampled the water, found it deficient, and asked my father to supply him with beer. This done, the work proceeded smoothly. Pomquist selected the stones he wanted along Carpenter Creek, and built quickly. He was a master stonemason.

1938: The following summer was one of much activity, most of it related to completing tasks the Nakusp carpenter had not been able to finish two years before. Dad cut out the windows and, reacting with horror to the size of the openings, rushed down to the ranch house to tell Mother that he had ruined the cabin. But when the window openings were framed with the broad planks specified in the carpenter's estimates, and fitted

The view from the cabin after Dad cut down forty trees, 1941. Pond family photo

with the sash Dad had ordered, doubts vanished. The windows were splendid. The Nakusp carpenter probably made the doors. Their particular construction is suggested in his estimates, and he probably found the blacksmith who made the hinges. Logs were chinked, and Dad built the log wall separating the living room from the rest of the cabin. My parents must have been immensely pleased. The cabin was habitable and beginning to be furnished.

1939: Susan was born this spring, and our parents came late to the ranch. Plumbing was completed, as were the kitchen cabinets.

1940: Dad built a terrace retaining wall to create a patio beyond the front door, and also an ice house. The plan was to cut ice on Loch Colin in winter, store it under sawdust in the ice house, and use it in summer in an ice box in the cabin. There would be no electricity in the cabin until 1964.

1941: With the cabin essentially built, Dad cut down forty trees to improve the view. Mother, presumably, had defended the birch clump in the middle of the accompanying picture. At the end of the summer, my father returned to Vancouver, and Mother, Ann (a helper), and my sister and I

Father roofing the ice house, 1940. Pond family photo

stayed on in the cabin well into October. We have Mother's letters to my father during these weeks.

Basically, these letters are preoccupied with two small children, my mother's uncertain health, and the question of how to support my grandparents when they could no longer look after the ranch house, let alone the ranch. I had tantrums—the worst when I insisted that "the" was spelled with an "a"—and one day fell out of a birch tree into Loch Colin (a dunking I remember). My sister was far more amiable and tractable. Mother had good and bad days, frequent stomach pains, and in one letter implied that she could have no more children. The problem of the ranch recurs throughout the letters: "I seem to leave every letter with a worry for you about the ranch, but I see the need every day and I know you are inclined to put things off! It seems to me the ranch is one of the toughest problems you have had to handle and I know you hate doing it, but I don't see who else will. It is one of the unpleasant duties of an eldest son."[1] There was talk, Mother reported, of hiring a man or a couple to look after the ranch and the ranch house, although it was not clear who was suitable or where he or they would stay. Granny and Granddad did not want strangers living with them. But they had agreed to close down part of the house for the winter, sell most of the cows, and treat the farm as a subsistence operation—an agreement that Sandy, who had been advocating these arrangements for years, treated with astonishment and Mother doubted would hold. She felt that she and her children should move down to the ranch house—Granny was lonely yet would not spend much time at the cabin—but Mother did not want to. She did not "live" down there. "Up here I do every minute. I love our own place more and more."[2]

Living room facing the front door, 1938. Pond family photo

1942–45: During these years, my parents spent little time at the cabin. To assist the war effort, in summer Dad cut firewood on the North Shore and worked as a welder's assistant in the shipyards. During their brief visits, the cabin's interior was becoming the furnished space my sister and I knew as we grew up. The furniture was all hand-me-downs, some of it from the Code family cottage on Lake of the Woods. My sister and I usually slept in the beds under the windows.

1946: Mr. Tamura, an elderly Japanese Canadian living in New Denver, laid flagstones on the terrace, reworked part of my father's terrace wall, and, using logs, large rocks, and gnarled roots, created a rock garden on the slope below the cabin. Hummingbirds returned year after year to nest in the lowest branch of the large birch on the patio.

1952: This year another artist visited the cabin, Sophie Atkinson, who at the time was living in Revelstoke, and whom Mother had got to know through her radio broadcasts. Sophie Atkinson was a tiny English lady with enormous energy who, as far as I could see, lived on lettuce and grapes. She was a watercolourist, and would set off each morning with a large easel and bundle of supplies altogether out of proportion to her

The furnished cabin my sister and I knew as children, ca. 1945. Pond family photo

size. She delighted in the cabin, painted it, and gave the painting to my parents—a fair measure of the building Mother and Dad had made.

AS I LOOK BACK ON THE CABIN NOW, MORE THAN SIXTY YEARS AFTER ITS COMpletion, I am struck by the extent of my mother's influence. Dad did much of the work, but neither he nor the ranch itself could have created a building like this. The very idea of it, drawn perhaps from summer cottages on the Lake of the Woods, was probably hers, and the building seems steeped in her aesthetic judgments. In her background in Winnipeg, as in my father's in Calne, was a wealthy, late Victorian grandparent, a product of business success. But unlike my mother's family's summer "cottages" on the Lake of the Woods, some of them very grand indeed, the cabin was not a monument to business achievement. Nor was it, as might have been expected on the ranch in the 1930s, a by-product of a social vision. I suspect, rather, that it emerged out of my mother's romantic imagination formed in the particular cultural and aesthetic climate of the 1920s, and intensified by her deep love of my father. She was creating their place with the materials and skills my father and a Nakusp carpenter provided, and making it as beautiful as she knew how.

That is as much as I can say.

Mr. Tamura's reworked terrace wall, 1946. Pond family photo

The Cabin, a watercolour painted by Sophie Atkinson in the summer of 1952. Pond family collection

8 THE JAPANESE CANADIANS

THE EARLY SLOCAN WAS A WHITE MINING CAMP WHERE "THINKING WHITE" was a common expression of approval. Miners feared that "cheap oriental labour" would undermine wages, and almost everyone shared the racism circulating in the late-nineteenth-century Anglo-American world. Very few Asians dared enter the early Slocan; those who did kept very low profiles. Among them before World War I was a succession of Chinese cooks on the Harris (Bosun) Ranch.

This was not the Harris family's first connection with Asia. The prosperous, upper-middle-class household in which my grandfather grew up in Wiltshire, England, was steeped in evangelical Christianity. Every morning in South Place, the family seat in Calne, family and servants assembled for prayers. Religion was pervasive, and, as I touched on in chapter 1, two of my great-aunts committed themselves to missionary work. One, Bessie, became a doctor (the Royal Free Hospital, London), and in 1892, under the auspices of the London Missionary Society (the evangelical wing of the Congregational Church), sailed for China where, well up the Yangtze River, she joined a medical mission in Hankow and married one of its doctors.

The other, Mary, a year younger than her sister and two years older than my grandfather, soon joined them. On shipboard, she met another missionary, and a year later married him. Eighteen days after the wedding, her husband died of acute amoebic dysentery. Mary was sustained by belief that his death was God's will, was somehow for the best, and that she and her husband would be reunited in heaven.

Six months later dysentery also claimed Mary. She had been in China for little more than two years, during which time the Christian missionaries in and around Hankow were decimated by disease and violence. Bessie and her husband returned to England in 1896, but had no intention of abandoning China. Bessie's father (my great-grandfather) in Calne wrote to his son-in-law:

> THE NEWS OF the death of Mr. Terrell preceded as it was by the news of the death of Dr. Turner, Mrs. Owen, David Hill [all British missionaries in Hankow], our dear daughter Mary and her husband Walford makes me and my wife and all our relatives and friends feel very concerned about Bessie and you, also dear little Walford [Bessie's infant son], we all feel that for Bessie to go to Hankow and to work there next summer is almost certain death to her, and very little less so to you, and we cannot believe that our heavenly Father desires that such a risk of valuable lives should be run.[1]

My great-gransfather had spoken to Bessie, found her heart "set on going," and asked her husband "to consider if you do right in subjecting your wife, yourself, and child to the climate of Hankow another summer—we all here, who love your wife yourself and child say No, a thousand times No." This appeal had no effect. Bessie and her husband returned to China. They became fluent in Chinese, loved the country and its people, and lived there into their old age. One of their children also became a medical missionary, and he and his family were in Hankow when Japanese troops took the city in 1938.

ON MAY 27, 1942, MY GRANDFATHER, J. C. HARRIS, WROTE TO HIS CHILDREN with the momentous news that the government was considering buying the Harris Ranch. It was wanted as a site for a sanatorium for "Japanese" relocated from the coast, and then as a convalescent home for returned men. The government official in charge, Mr. Boultbee, had visited the ranch, asked if it might be leased or sold, and explained that the government would prefer to purchase "as they would have to spend a good deal of money to fix it up and would not want to improve other people's property." Dr. Francis, the local doctor, strongly favoured the sanatorium. "He says it [the ranch] is an ideal spot. Centrally situated, fairly close to town, but not too close. He also thinks that it is high time that both Granny and I retired as the load is far too heavy for us, which is true." The government could expropriate at a price set by arbitration. A decision was imminent: there was little time to get those being relocated, all of Japanese ancestry, settled in and a garden planted. The deal might not go through, but if an offer came my grandfather told his children that "we ought to sell ... the place will deteriorate rapidly and I cannot keep it up. We are too isolated for old folks, if we are not feeling very well it is too hard to get help, though Sandy and Mollie have been very good indeed."[2]

The whole matter, Granddad wrote, was a huge shock. "I feel sorry for you all. Sandy and Mollie are also upset, they love their very nice cottage ... Every one of you have made great contributions to the old home. Marie [widow of Tod, the youngest Harris son, who was killed by lightning] has not had much chance but we think of her as quite one of us. We must stay united in spirit whatever happens." "What a home," he wrote, "this has been—Granny's work of art—and a masterpiece."[3] In the shaky hand of a Parkinson's sufferer, Granny added a note to my mother: "It is sad to think that our happy summers at the dear old Ranche are at an end. There may be good times there again, but it will never be the same again."[4]

The negotiations with the Security Commission (the body mandated by the federal government to oversee the removal and resettlement) are not known, but their outcome is clear. In August 1942, my grandfather leased the ranch and its buildings to the federal government for fifty dollars a month "for the duration of the war with Japan," and also gave the government an option to buy the north forty acres of Lot 1800 for three thousand dollars.[5] My grandparents moved into the Bosun Mine manager's old house. Sandy and Mollie were allowed to remain in their house and my parents to use their summer cabin above Loch Colin. The

Sanatorium at New Denver, ca. 1943. The sanatorium is the large building fronting the beach; the small houses in "the Orchard" behind comprise the principal settlement for Japanese Canadians in the Slocan. Betty Andrews and Midge Ayukawa Collection, Nikkei National Museum, 1994.62.48

sanatorium was built on the shore of Slocan Lake beside the internment camp in New Denver (the Orchard), not on the Harris Ranch.

According to my father, Richard "Dick" Harris, his father cooperated with the Security Commission partly because he was aging and the ranch had become a burden, and partly because he was "sorry for the evacuees and anxious to make their lot less arduous in any way he could."[6] This is probably accurate, but my grandfather did consider that the evacuation of the Japanese Canadians from the coast was necessary, and explained why in a short essay about their coming to the Slocan.[7] Anti-Asiatic feeling, endemic in British Columbia for years, had led, he held, to unjust laws, acts of violence, the segregation of Asiatic populations in coastal cities, and, on one occasion, "the forcible prevention of a shipload of Hindoos from landing in Vancouver."[8] Behind this prejudice, he thought, was the fear of cheap labour.[9] "The white workingman felt that these experts in the art of frugal living would be used to undermine his standards. The small business man also became aware of the 'Yellow Peril' as all sorts of small shops and trades were opened with Asiatic ownership." For years prejudice was vented against all Asians, but when Japan attacked China in 1937 public indignation turned particularly against the Japanese, and with mounting fury after Pearl Harbor.[10] People were afraid. The Japanese might attack by sea; Japanese fishermen and their boats were already on the coast. There was danger of mob violence against isolated "Japanese."[11] My grandfather thought that in such circumstances the provincial and federal governments had to take quick, drastic action, and decided "to move all the Japanese to some places in the Interior where they could be watched and guarded." The Slocan Valley, isolated but accessible by branch railroads, lacking important targets for sabotage, containing many abandoned buildings in fair condition (lumber was extremely scarce), and well forested, was seen as an obvious place for them.

The quick relocation of most of the some twenty-two thousand people of Japanese ancestry in British Columbia was, my grandfather thought, a "gigantic task" that "from what we were able to see" the Security Commission accomplished "with great credit and humanity to the Japanese."[12] There was a need, he thought, for speed, first to assemble them in Vancouver, then to get them away from Vancouver "where conditions were very bad" and into camps in the interior. Moreover, many interior communities were wary of receiving large numbers of people they knew nothing about and were predisposed to dislike and fear. The commissioners consulted locally, but "let it be known that they had to act promptly and that no private interest could prevail against their decisions." Essentially, my grandfather agreed, and thought that for all the mistakes and hardships, the Security Commission had handled the situation well and with commendable consideration for people caught in circumstances beyond their control.

Some forty Japanese Canadian carpenters (under white foremen) were sent to Sandon and Kaslo to repair unused buildings, and contingents of evacuees soon arrived to occupy them. Other carpenters went to Slocan, there to be closely followed by residents—"almost too closely," wrote my grandfather, "for there was terrible crowding for a time." In New Denver, the situation was different. There were few abandoned buildings to fix up; rather, a camp would have to be made, and there was a good deal of local opposition to doing so. New Denver was a village of some 350 Caucasians, and the prospect of suddenly accommodating several times as many Japanese Canadians, when most people in New Denver had never seen a Japanese person and Canada was at war with Japan, was easily terrifying.

Eventually, however, the view prevailed that the Japanese Canadians had to go somewhere and that New Denver should bracket its prejudices and make a virtue of necessity. The New Denver Village Council and the Board of Trade voted to accept them. The village of Silverton did not. The management of the Mammoth Mine feared that the company's concentrating mill might be sabotaged, but anti-Japanese feeling in Silverton ran well beyond mine managers. When the Silverton Miners Union obtained permission to bring two Japanese Canadian baseball teams to play there, so much bitterness developed that the plan was dropped.

At first the Japanese Canadians trickled into or through New Denver: one or two *en route* between Slocan and Sandon, others sent from Sandon to the New Denver Hospital. Dr. Francis hired an old Japanese Canadian fisherman to work in his garden ("we all admired his industry and the fine care that he took of the garden," wrote my grandfather. "He was an Anglican"[13]). In September, Japanese Canadian nurses trained in British Columbia began to work in the hospital ("they made a most excellent impression with the patients, and with all with whom they have come in contact. I think the nurses have done very much to heal both the sore bodies and the sore and bitter feelings between the whites and the Japanese. No one could help liking them."). At this time, building began in earnest. The Security Commission rented the skating rink, which was roofed and in good condition, and many newly arrived evacuees stayed there for a time. Part of the rink was turned into a carpenter shop and fitted out with power tools; there carpenters made the prefabricated components of huts they would then erect in a few hours. By the end of January 1943, some 1,500 Japanese Canadians were living in and around New Denver,[14] most in the small standardized huts common in most of the camps, but some still in tents.

Japanese evacuees find themselves in a new settlement in the Slocan area, 1942. The basic camp house is in the foreground; the tent behind is of the type in which some families spent much of the first winter, one of the coldest on record. Leonard Frank photo, Alex Eastwood Collection, Nikkei National Museum, 1994.69.4.16

The first Japanese Canadians to arrive on the ranch—two men and the rest women and children—were trucked over from Kaslo in an open two-ton truck on a chilly day in late October or early November 1942.[15] No one expected them. The twenty-five newly built houses in the Far Field had stoves but neither stovepipes nor firewood. My uncle Sandy gave them cocoa and rustled up some stovepipes and wood. The newcomers crammed that first night into a few rooms, but then, when basic provisions arrived, began to disperse into the small houses, exactly like those in New Denver, that had been erected in the Far Field. Other families came later from Kaslo, or from the Orchard in New Denver when some huts there were converted into a school.[16] A few families went to the ranch house, but left when the commission decided to make it into an old men's home. Apparently, some of the old men came from Sandon, where they had been lodged in the Sandon Hotel. There, as they told a commissioner, they had "become use to the climate and environment" and were determined to stay: "No the old men are animated easily and become so unrest with fear feelings so that by not means cannot obey your orders."[17] Probably, however, they were moved. At the height of the Japanese Canadian occupation, some fifty old men and two cooks were in the ranch house and between 150 and 200 men, women, and children were in the houses on the Far Field.[18]

Harris Ranch Old Man Camp, New Denver, 1945. This was the ranch house, home to some fifty elderly men and two cooks. The small, individual flower gardens that many of them made were at the back of the house. Izumi Family Collection, J. T. Izumi photographs, Nikkei National Museum, 2012.29.2.2.23

My grandfather considered the winter of 1942–43 "the hardest we have ever experienced in the Slocan," and thought conditions in the camp "awful for a time." As soon as fires were lit, the green hemlock planking that had been used on the walls (no other wood was available) shrank and sweated. The huts were wet and drafty, and with wartime scarcities there was no material to fix them. Moreover, firewood remained in short supply. Work crews composed of men from the camps salvaged hundreds of cords of dry firewood from burned-over land near Summit Lake, and trucked some of them as far as Sandon. Some of this wood may have reached the ranch, but with forests at hand the evacuees on the Harris Ranch soon cut most of their wood there. The ranch house had a ravenous furnace and hot water heating system, and an adjacent water-powered sawmill with a circular saw set up to cut logs into stove and furnace wood. The old men used it that first winter while struggling to find dry wood. They also struggled with an unfamiliar heating system, let it freeze, then ran stovepipes through the house's many rooms. While life on the ranch in the winter of 1942–43 must have been a struggle, the impression made on the Harris

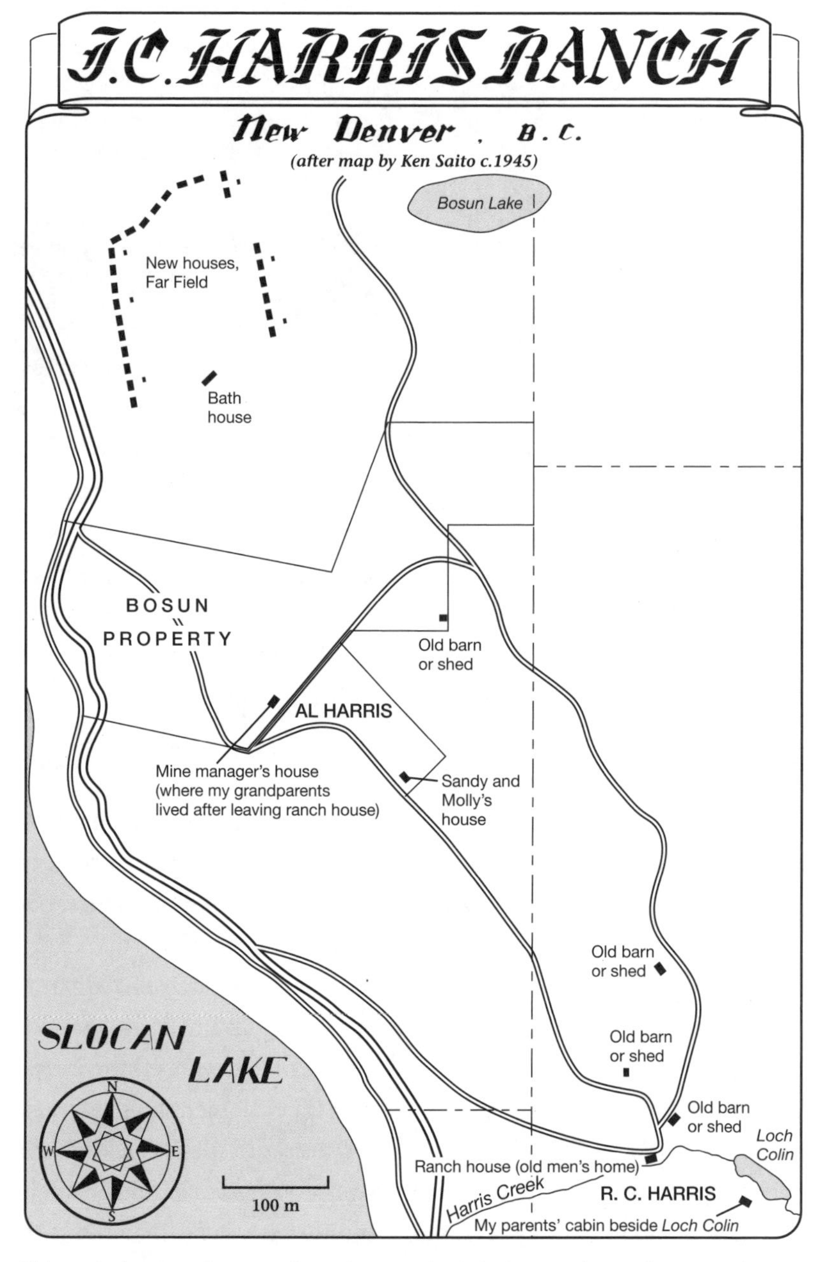

Although drawn only approximately to scale, Saito's map shows the general arrangement of roads and buildings. It does not show that all the buildings were on a narrow bench bounded above and below by steep, forested slopes. J. C. Harris Ranch, New Denver, ca. 1945, modified from a map by Ken Saito.

members of the ranch was of the resilience of people who had reason to be sullen and angry. "We felt," wrote my grandfather, that "our new neighbours deserved great credit for the plucky and cheerful way in which they met their hardships."[19]

Leaving the ranch house for the mine manager's house on the Bosun Mine property in the fall of 1942 was emotionally wrenching for my grandparents, particularly for my grandmother, and she did not return, even for a look, until early the following July. The next day she wrote to my mother:

> I DO FEEL that everything has gone "topsy turvy." The good old Ranch days seem to be over altho I have never seen it [the ranch house] look more beautiful than it does now. I went to see it yesterday and never in my best days did I have such a flower garden. I asked for some flowers to take to the cemetery and got an arm full. The rows and rows of vegetables certainly look splendid. I didn't go into the house, it was very tidy and neat outside, more so than it often was in my time.[20]

It was the difference between two elderly people struggling to maintain a house and gardens, and fifty old men with time on their hands.[21]

By this time I was a small boy, and I remember the old men in the ranch house. I remember them chatting on benches along the porch, cutting firewood in the small sawmill just across the creek on the road toward the cabin—and the stories my uncle Sandy told about the two who each lost an arm there. I remember them heading up the bank, into the forest, with empty flour sacks, and returning, their sacks bulging with mushrooms. And the huckleberry pies. There were lots of huckleberries in those days, and when we mounted a day-long, family huckleberry expedition, commonly up the North Fork (Kane Creek) of Carpenter Creek, we usually filled several large boxes. One box always went to the white-aproned cooks in the ranch house from whom, in return, we got a couple of pies. Above all, I remember the gardens along the little creek behind the ranch house.

Old men and cook (in white apron) on garden steps in front of the ranch house, Harris Ranch, ca. 1944. Note the flowers beyond the fence in what had been my grandmother's garden and, in the background, some of the forty cords of wood required to heat the ranch house in winter. Harris family photo

The old men divided the banks of the creek into small plots, each no bigger than the area of a double bed, and turned them, one by one, into a mosaic of ornamental gardens. They scoured the shores of Slocan Lake for wave-worn rocks and twisted, weathered roots. They built small water wheels that turned in the creek, and bridges, one of which was big enough to walk across. They cut patches of grass with scissors, and planted flowers. Theirs was, I think, a competition of small gardens, few of which, I suspect, had much to do with either traditional Japanese or Western gardens. They were probably the hybrid creations of uprooted people. Whatever they were, they were dazzling. Even as a kid more interested in play and hunting frogs than in gardens, I admired them. When the old men left, their gardens were soon overgrown; their remainder to this day is a tough, spiky, feral grass.

The people in the Far Field were farther away, and when my sister and I and our parents were in the cabin near Loch Colin we saw little of them. I may have played a little with the boys there, but my memory is hazy. My cousin Nancy, several years older than I and living close to the new houses in the Far Field, knew the children on the Far Field far better than I did. The ranch was an attractive place for children, and memories of childhoods there, as often in other camps, are relatively positive: skating, skiing, and sledding in winter; swimming, fishing (even for shiners in Bosun Lake), and a mountainside to explore in summer.[22] For their parents, life was undoubtedly tougher, partly because stores in New Denver were more than a mile's walk away,[23] partly because of the crowding common to all the camps, and partly because of the challenges of working out social relations among people awkwardly thrown together. When, for example, the Japanese Canadians built a bathhouse on the Far Field, the men bathed every night; the women complained, and only, so the Harris family understood, after much negotiation within a traditionally patriarchal society was it agreed that men and women would bathe on alternate nights. There are not, apparently, written records of life on the Far Field, but there are photographs: a mother, her baby, and a planting of flowers in front of one of the houses; my cousin Nancy with skiers; two little girls, one of them wearing a pair of patched pants that, according to her older sister, was a Harris hand-me-down that had once been mine.[24]

A portrait of Ume holding Mezumi at Harris Ranch, New Denver, in 1944. The typical evacuation house is in the background, its aspect considerably softened by a fine showing of Japanese anemones and a well-made trellis for beans. Izumi Family Collection, J. T. Izumi photographs, Nikkei National Museum, 2012.29.2.2.25

There are also pictures of the crescent of small, shake-roofed houses overlooking Slocan Lake, one of them with accompanying Japanese text.

Two little girls, Harris Ranch, 1944. According to Sachi Manuel, who kindly provided this photo, her younger sister (left) was wearing Harris family hand-me-down overalls that I had once worn. I don't remember them! Manuel family photo

That text, which might be assumed to deal with the hardships of an internment camp, turns out to be a poem, which my friend Tsuneko Kokubo has translated as follows:

Lake kingdom, New Denver
The images of mountains remain
as they were
On the hillsides white clouds,
summer still cold
Mist, the fragrance of blossoms
and green leaves

Of all the members of the Harris family during these years, Uncle Sandy was most involved with these newcomers to the Harris Ranch. For one thing, he and Mollie had a phone, and whenever there was an emergency—an arm cut in the sawmill, a baby arriving unexpectedly—someone informed Sandy. He also had a radio, which the evacuees were not allowed. Mrs. Tomi Fujihara, who looked after my grandmother during her last months, was having dinner with Mollie and Sandy when the news of the atomic bombing of Hiroshima came over the radio. "Oh oh," said Mrs. Fujihara quietly, "that's where my two children are" (her son, she later learned, was killed, her daughter badly burned).

Sandy liked the evacuees—the men in the ranch house were "a dashed nice bunch of old guys"[25]—and they him. Takayuki "Tak" Hatanaka, who became his right-hand man in the New Denver Light and Power Company, was a particular friend. In one story Sandy loved to tell, the company was needed in Silverton where the evacuees were not allowed. After negotiations with the village council and also the manager of the Mammoth Mine, who was particularly hostile to the "Japanese," Sandy obtained permission to bring Tak to Silverton. They had not been at work long when a lad came to say that the school flagpole was "haywire" and the Canadian flag was stuck thirty-five feet up. Could they come and fix it? "Tak," said Sandy, "how would you like to climb that flagpole?" He said "sure" and went up and fixed the Canadian flag.[26]

After the war most of the evacuees in British Columbia were offered the choice, which had to be quickly made, of moving east of the Rockies or of going to Japan. All those on the Far Field faced this choice, made it, and soon left. Their houses were demolished. The old men in the ranch house could remain in the province, and apparently many of them moved to Slocan City. A Mr. Tanaka stayed—"Mr. Sandy, I don't want to go. May I stay in this house?" From time to time he worked, unasked, in Sandy's garden. Sandy tried to pay him, but failed, and, noting that even on the hottest days the old man worked in rubber boots, bought him a leather pair. "Thank you, thank you, Mr. Sandy," said Tanaka, but he never wore them. Months later, he invited Sandy to his room for a drink of homemade sake. "Mr. Sandy, come, drunk." At the head of the bed, arranged on a shelf, was a long, neat row of boots, Sandy's at the end. When Mr. Tanaka died in 1951, the Japanese Canadian years on the Harris Ranch were over. The ranch house sat empty, quickly deteriorated, and was taken over by pack rats.

In August 1944, Prime Minister Mackenzie King addressed parliament about the future of the "Japanese" in Canada, and alluded to "the extreme difficulty" of assimilating them into Canadian society.[27] This was not the view from the Harris Ranch. In a letter to the *Nelson Daily News* in August 1943, my grandfather wrote that the Japanese Canadians had "many noble qualities that [would] eventually enrich Canadian life."[28] A year later, and in the same vein: "It will be a great reproach to ourselves if we fail to turn these very gifted and energetic people into very important ingredients of our future Canada."[29] In February 1946, he wrote to Mackenzie King about New Denver's experience with the evacuees: "We tried from the first to be helpful and friendly to them and we won each other's esteem and respect. The oft-made assertion that The Brown Man

My cousin Nancy with young skiers on Harris Ranch, 1944. With skiing, sledding, and skating in winter, and swimming, fishing, and games in summer, the ranch seems to have been an attractive place for young evacuees. Given the distance to shops in New Denver and the myriad disruptions of relocation, life at the ranch was far harder for their parents. Harris family photo

and The Whites will never mix has little foundation. The children play together in most friendly fashion. The young people dance together."[30] In June 1949, less than two years before he died, he wrote to his friend Mrs. Hoshino, who had lived on the ranch during the war:

> I WAS VERY pleased indeed when you told me that you intended to take out your naturalization papers, for I think that since fate has made your home in Canada, that it is right for you to join in with us to the full, in our national life. We had the pleasure of seeing a great deal of you and Mr. Hoshino and your little boy and girl when we were close neighbours, and we liked you all and I felt that you were raising two nicely behaved children who will grow up to be a credit to their parents and to Japan and to Canada.
>
> Those were happy times up on the old ranch, though clouded by the deaths of your husband and of my wife. May you

Far Field houses, ca. 1944. This photo seems to have been taken early in the Japanese Canadians' time on the Far Field, before gardens were established or the bathhouse built. A006-000-0160, Silverton Archives

live long and happily in Canada and your children develop to great Canadians.

From your old friend, J. C. Harris[31]

Presumably Mrs. Hoshino's view of life on the ranch was much more qualified than my grandfather's, but she lived on in New Denver and remained a Harris family friend to the end of her long life.

THROUGH THESE SAME YEARS, THE HARRIS COUSINS WHO HAD BEEN MEDICAL missionaries in Hankow were interned in a Japanese camp for prisoners of war. Immediately after Pearl Harbor and, the next day, the Japanese invasion of Hong Kong, Japanese soldiers in Hankow confiscated foreigners' radios and cameras and curtailed their mobility. Medical missionaries and their families required permission to leave the hospital compound. A few months later, the army took over the hospital; foreign enemies of Japan were to be sent to Shanghai, either to be repatriated in exchange for Japanese prisoners of war or sent to "Civil Assembly Centres" (internment camps). In August 1942, the Harris cousins (Bessie, her husband, and their eight-year-old son, Walford) and other foreigners were loaded on a gunboat and sent downriver.[32] A few Americans were repatriated, but the others—mostly British plus a few Americans, Dutch, and Belgians—were crammed for several months into a temporary holding station in Shanghai, then sent to an internment camp in Yangchow (Yangzhou), almost two hundred miles to the northwest.

On their arrival, the camp commandant informed them that the emperor of Japan had kindly chosen the camp for them, and that they would be shot should they try to escape. The camp itself had been an American missionary boarding school, and several families were jammed into most of its many rooms. The Harris cousins were three of thirty sleeping in a room of narrow cots separated by hanging straw mats. There was no privacy; whatever one said or did, others heard and knew. The daily timetable never varied: wake up at 6:30 a.m.; breakfast from 7:00 to 7:30; muster and roll call in the yard at 8:30, at the end of the day; and lights out at 9:00 p.m. Food was minimal and everyone was always hungry: some rice, some flour (the internees made bread), sometimes a little pork, sometimes a few vegetables, and occasionally food parcels from the Swiss or American Red Cross. A watery vegetable stew, at times with a little pork, was a staple. Barrelled water was brought in wheelbarrows and stored in tanks (*kongs*) to let sediments settle, or obtained from wells the internees dug. All water had to be boiled, and the scarcity of wood limited its availability. The daily water issue per family for all uses was two or three Thermoses in winter, one in summer.

As the war turned against Japan, conditions in the camp worsened. Food parcels became rarer, then stopped. Some of the Japanese soldiers became vindictive. One of them, Tanaka, was particularly disliked and brutal, often forcing the lines of prisoners in the morning musters to stand in the full summer sun until some fainted. On one occasion he forbade the use of wood for boiling water, and on another beat an *owai-ya*—one of the peasant women who came into the compound each morning to collect human waste (night soil)—within an inch of her life for claiming that the war was over. The Korean guards, on the other hand, were thought gentler and friendlier; occasionally they played with camp children. Commandant Hashizumi, who had been himself a prisoner of war until repatriated in a prisoner exchange and was in charge during the camp's last months, apparently did what he could. He allowed prisoners to keep a few goats and plant vegetable gardens and, with extraordinary courage in the face of imperial edict, apparently dipped into military funds for food. Even so, conditions in the camp during the last months of the war were desperate. Life was hanging by a thread. The children were taught, there was a lot of campfire singing (without a campfire), but the staples of life were barely at hand. Children received a cup of goat's milk every ten days; the water ration became a cup per person per day. Starvation was averted when the Swiss Consulate in Shanghai provided funds, enabling the purchase of some pork, rice, and sweet potatoes.

IT IS EASY NOW TO JUDGE, BUT DECISIONS AND ACTIONS IN THE PAST SHOULD be contextualized in the thought and emotions of their time. Given the war, the fears, the intensity of anti-Japanese feeling, and reasonable doubts about the loyalty of some, my grandfather may have been right that the government had little choice but to remove people of Japanese background from the coast, and to do so quickly. Certainly, he was right that the relocation was a huge logistical challenge, and perhaps even right that, for all the hardships, it was accomplished with a fair measure of humanity. There were no machine guns, barbed wire, or soldiers in New Denver, nor even, eventually, the extra police the Security Commission had intended. In awkward new circumstances, the evacuees were reassembling their lives, and as time passed and people got to know each other, the relations between them and the prior New Denver population became increasingly cordial. The New Denver baseball team began to win, the young to mix at the dances. Over the years, the camps have been variously remembered, sometimes angrily, sometimes not. There is no doubt that when the evacuees arrived conditions were very bad, no doubt that lives were drastically altered, yet there is a sense that the camps in and around New Denver were also creative places where lives were being worked out in new ways. One might even say that something of a subsequent Canada was beginning to come into focus.

The basic decisions—to remove Japanese Canadians from the coast, to confiscate and then sell their property, and at the end of the war, to force most of them to choose quickly between relocating east of the Rockies or returning to Japan—were variously rationalized. Relocation, it was held, was a legitimate and necessary response to a fear of Japanese invasion, to doubts about the loyalty of some of the Japanese Canadians, and, given the intensity of anti-Japanese feeling, to concern about the safety of small, isolated populations of ethnic Japanese. Forced sales of fish boats were justified because, unused and deteriorating, they were wasting assets (officials initially intended sales to be between owners and buyers); it was even argued that after the war the government might be sued for mismanagement were boats not sold while they still had value. Sales of landed property, particularly of unused buildings and abandoned farmland, were similarly, if less convincingly, justified.[33] Officials held that the government, faced with the myriad demands of war, did not have the administrative capacity to look after many hundreds of confiscated properties. After the war, the abrupt relocation of most Japanese Canadians east of the Rockies was treated as a means of removing them from ethnic ghettoes and assimilating them into a larger Canadian society.

A good measure of racism lurked behind such rationalizations. From its colonial beginnings, British Columbia thought of itself as a British settler colony, although its location on the Pacific Ocean some 7,500 miles from Britain tended to destabilize this ambition. A desired future was not secure and, because it was not, the racism that was endemic in the late-nineteenth-century Anglo-American world was particularly strident in the newly established colonial society of British Columbia. For the most part, this was a generalized anti-orientalism, but in the late 1930s, and particularly after Pearl Harbor, Hong Kong, and the Philippines, it focused on the Japanese Canadians. The fact of the matter was that a large majority of British Columbians did not want them in their province. Often they said so. When they made other arguments, the common, unstated subtext was a racialized denial of a place in British Columbia for Japanese Canadians.[34]

Mackenzie King admitted as much when he reported to parliament about the government's plans to relocate the Japanese Canadians after the war.[35] He had received, he told the House, many representations to keep the "Japanese" away from the coast, and argued that in the interest of racial harmony they should be dispersed across the country. So distributed, they would assimilate into Canadian society; concentrated, as he was assured they had been in British Columbia, they generated racial hostility. Canada was fighting a war against fascism and racism, and to prevent the growth of racism in Canada he argued that the Japanese Canadians had to be moved out of the province and dispersed.

King was not entirely comfortable with this argument. He acknowledged that the behaviour of the Japanese Canadians in the camps during the war had been exemplary. Nor did he think it good policy to prevent citizens from moving freely within their country, and held, therefore, that the forced dispersion of the Japanese Canadians would be temporary. But he had to deal with the prevailing British Columbian "dislike, fear, and distrust" of the "Japanese"—with white, British Columbian racism[36]—and did so by reframing the issue as a Canadian rather than a British Columbian "problem" that required a Canadian solution. The "Japanese" were being dispersed across the country with the high-minded purpose of thwarting the growth of Canadian racism—or, to make the case with less political spin, to curb the racism of white British Columbians.

Throughout the Slocan's early modern history, the overwhelming majority of its people shared these racist views, and with a particular edge of fear that "cheap oriental labour" would undermine the hard-won gains, fought out in labour disputes throughout the North American West, of white working men. As a result, as noted earlier, almost no people of any Asian background ever entered the Slocan, and the few who did kept exceedingly low profiles; the Chinese cook on the CPR train to Sandon would not leave the train for fear of his life. Yet, the New Denver Village Council and Board of Trade agreed to admit the Japanese British Columbians, partly, apparently, because of the persuasions of Dr. Francis, an influential voice in the village, and partly, one suspects, because members of the Board of Trade in a sleepy Kootenay village thought they would be good for business. Yet there was much apprehension, even fear: Canada was at war with Japan, these were people of Japanese ancestry, they were not white, and, when established in the camps, they would be five times as numerous as the prior, white population. The Slocan seemed ripe for misunderstanding, tension, and ill will.

Yet, it seems clear that such reactions were fairly rare. Fears tended to dissipate, the Japanese Canadians were increasingly liked, and there was surprisingly little racial tension. Certainly, there were grumblings; some prior residents did not want these newcomers around, but there were many others, my Slocan relatives among them, who admired and lived on friendly terms with many of the evacuees. For their part, the Japanese Canadians had ample grounds to be bitter: the majority were Canadian citizens yet their properties had been confiscated and they had been treated as enemy aliens. Given their backgrounds, their values, and the war, both prior New Denverites and interned newcomers had grounds to be apprehensive and hostile. Yet such values do not seem to have dominated New Denver during the years when Japanese Canadians were its principal population.

The conduct of the evacuees themselves must be a good part of the explanation. From the first to arrive—Dr. Francis's gardener, the nurses in the hospital—the evacuees' politeness, apparent calm, competence, and fortitude were quietly impressive. A resilient people had been dropped into the Slocan. A good many of them probably reacted to their changed circumstances much as did Tatsue Nakatsuka, who became a teacher in Sandon: "My initial dismayed reaction to Sandon had gradually changed into a much warmer feeling. I think I acquired a lot of wisdom

Japanese Canadian employees, Greer's store, ca. 1944. Many Japanese Canadians were quickly integrated into the larger life of New Denver. Even the New Denver baseball team, its record dismal for years, suddenly began to win. New Denver Museum and Archives, 2001.029.007

from the whole experience. I learned that adverse conditions can be overcome. It all depends on how one wants to accept things. The only difference is whether one wants to work with the flow of things or against it."[37] A similar attitude perhaps explains the nature poem on the photo of the camp in the Far Field (see page 165), and the quilt of flower gardens around the ranch house. I am told that the common Japanese term *shikataga nai* (literally, "nothing can be done") alludes to what seems to have been going on here—a willingness to accept circumstances beyond one's capacity to change. White leadership must also have been important: Dr. Francis was always an advocate for the Japanese Canadians, as was my grandfather, especially in his role as chair of the local committee on community–Japanese relations. Beyond leadership was probably the more basic fact that, as lives increasingly overlapped, a good many of the older population of New Denver and a good many of the Japanese Canadians were coming to know and appreciate each other.

To be sure, the local geography of resettlement was segregated. The New Denver delta north of Carpenter Creek was Caucasian, the land south of the creek (the Orchard) Japanese Canadian. In most of the smaller, neighbouring camps—Thrings Ranch, Nelson Ranch, Rosebery, Harris Ranch—Japanese Canadians made up most of population. In all the camps, many people, particularly the elderly, spoke no English and had little or nothing to do with English speakers. Similarly, many of the prior, white population did not interact with the newcomers. Yet there were many points of interracial contact: in the stores, where many shopped and young Japanese Canadians were frequently employed; in a variety of local jobs; in the various Christian churches, which many of the Japanese Canadians began to attend; in the hospital; on teams and playing fields; at dances; with white teachers in some schoolrooms; at theatrical performances; and in countless informal encounters. Some of these contacts were glancing, but others, as the evidence from the Harris Ranch suggests, were close and deep.

How, then, to view the Harris Ranch during the years when the Japanese Canadians were its principal population? The evacuation, a product of war and racism, is not a bright moment in the Canadian past, but it and the camps that followed are understandable by-products of their time. In some respects they are a measure of the failure of white British Columbians to come to terms with diversity, but something else, something creative, does seem to have been stirring on the Harris Ranch, and probably in other camps where Japanese and other Canadians mixed. People of very different backgrounds who had come to British Columbia from opposite directions and across different oceans were learning how to appreciate and live with each other. There was not much time, three or four years, before lives were uprooted again, but these years seem to me to provide a glimpse of the Canada that would emerge more fully as Japanese Canadians were allowed to return to British Columbia (1949), immigration laws were transformed (1966), and it became increasingly clear that the appreciation and accommodation of a good measure of diversity were built into the nature of Canada.

9 DEATH AND SUCCESSION

ON MARCH 29, 1951, JOSEPH COLEBROOK HARRIS DIED IN HIS SLEEP AT HIS brother's house in Victoria. He was eighty years old. For some time, his heart had been slow, around thirty beats per minute, and that night it stopped.[1] Arrangements were made to send his body to New Denver, and for a memorial service in Turner Memorial United Church. My father and aunt Heather would drive to New Denver for this service, and I, then fourteen, would accompany them. The drive was long and the road between Christina Lake and Rossland almost impassable. Its surface was thawing and oozing downhill. We got through because we were between Greyhound buses; at the worst spot a tractor pulled us through. I don't remember any details of the memorial service, but I do remember that the small church was not full and that neither the service nor my grandfather's death had much impact on me. I had seen him off and on all my life and had enjoyed his affectionate boisterousness, but in any larger sense I hardly knew him. Nor, at fourteen, had I much sense of death.

My grandfather left two wills. In the first, signed in December 1940, he left all his property, except the ranch itself, to be divided among his four children. The ranch was another matter because, in my grandfather's view, there was no easy way of dividing it:

> [THE RANCH ITSELF] could only be divided in my opinion with great difficulty and expense and would be much damaged if divided up. Water rights & rights of way to make each part accessible would be involved. However this is a matter which the four trustees must decide to best suit themselves. I only desire that the old

> ranche which has been such a very happy home for us all shall continue as the family centre and be at least a frequent resort for our grandchildren.
>
> Alexander Leslie [Sandy] & Richard Colebrook [my father] have by the valuable improvements that they made established especial claims to the homes that they have built on this property but with no legal standing. This has been possible because we have maintained such friendly relations and confidence in the good faith of each other. I desire this state of affairs to continue and express my equal love and confidence in you all and thank you for your constant kindness to your mother and myself.

Four years later, he replaced this will with another that left all his "real and personal estate," including the ranch, in equal shares to his children, their spouses, and his grandchildren. The ranch, he again acknowledged, "will present difficulties in division so as to preserve its usefulness," but held that "by consultation & cooperation I trust that you will arrive at just and satisfactory arrangements." His four executors—his three surviving children and Marie, Tod's widow—were given full powers to carry out the purposes of the will.

Although a vivid reflection of my grandfather's values and of his trust in his children, his will poorly fit his children's different relationships with the ranch. Basically, my grandfather's will was symmetrical and egalitarian, whereas his children's relationships with the ranch were not. Sandy had stayed and the others had left. After his months there in 1924, my father had moved to Vancouver and become a high school teacher. Aunt Heather had married a mining engineer and lived in the Bridge River Valley. Before his death, Uncle Tod worked for the Hedley–Mascot Mine and lived in Hedley in the Similkameen Valley. All of them returned to the ranch as often as they could, but none of them lived there. Sandy and Mollie did. Moreover, for many years Sandy had done much of the practical ranch work, probably with relatively little appreciation from a father who was not good with equipment and whose thought turned toward social reform. My father, who was more bookish than Sandy, may have been the more favoured son. If so, and from Sandy's perspective, my grandfather's will would have been yet more evidence that his talents and work on the ranch were undervalued. In Sandy's mind, the ranch should have been left to him.

Early in 1953 the Vancouver law office handling the will urged the executors to work out the solution for the ranch that my grandfather had left to their best judgment. Sandy twice offered to purchase the ranch, offers that were rejected apparently because my mother and father were committed to the cabin and its surroundings and Aunt Heather thought that her children might want some lakeshore. The ranch could have been divided to accommodate these interests but, surrounded as it was by affection and emotion, the family could never bring itself to discuss the options. Granddad's will was not addressed, the ownership of the ranch remained much divided, and, when he allowed himself to do so, Sandy became thoroughly fed up.

His feelings emerge occasionally in his weekly letters to Nancy, his daughter. *March 14, 1953*: There was, he told Nancy, too much ranch work, he could hardly do it, but "hated to see the place going to the devil." The R. C. Harrises were not really interested in the ranch, and hardly deserved such a place. "The way this place and my time have been mucked up makes me pretty darn mad when I think of it." *June 6, 1954*: "The big trouble is that most of the family don't understand the Ranch problems in the slightest way and they don't understand that they don't ..." *Feb. 23, 1957*: The only way to stop worrying about the ranch and all the work put into it, "is to get away from this place, and somehow I think Mums and I would be very miserable if we did that. We have been here too long, and got too involved to be able to walk away from it. On the other hand, it's stupid to go on the way we are doing. The aggravating part to me is that the Family don't seem to have a clue as to what they want to do with the place."

The question of the ranch was still unresolved when, early in 1959, Heather, Sandy's sister, wrote to him about it.[2] She pointed out that their father "couldn't have left a more complicated situation—nor a more dangerous set-up for blasting a family apart," that the family was precious, and that "back of that family is the Ranch and all it means." She hoped that "this will business" could soon be settled.

Heather's views were not her brother's. She thought that Sandy had been given a good deal—a home on the ranch, the New Denver Water Works—and that if he wanted the ranch he should buy it. She asked Sandy to renew his earlier offers to buy, stating the amount of property he would like to own. "Send a copy to Dick and one to me. Then if and when we can come to some terms—lets get a surveyor to complete that part." Her children, she thought, might have some interest in campsites along the lakeshore, not more. In a note to my father, she hoped Sandy would find her letter sincere, as it was. She knew that "the whole thing upsets him—but I wish he'd realize how very insulting and upsetting he can be."

Sandy did not reply. In June his daughter, Nancy, saying that she was "very tired of hearing about all the ranch problems which seem to be no nearer settlement than they were ten years ago," wrote to her aunt.[3] Her interests in the ranch, she said, were sentimental, as were her father's. "Had he not had some feeling for the place he would have left it long ago and the problem of spending a life time of work on something that was not his own would not be so disturbing to him." Few would have worked as hard on the ranch as her father "for no return at all but purely for the satisfaction of the work and the results." She thought that if the ranch were used as a ranch, then one person should own and operate it; if used for something else, that should be decided upon so her father knew where he stood. The lakefront, she thought, was unsuitable for cottages. Moreover, her father had received legal advice that any plans to subdivide the ranch would be complicated, and he "refuses to have any agreement that will mean more long legal complications." She hoped for a concrete agreement soon, was afraid that hard feelings were creeping in, and thought it "very unfair to the person whose problems of the ranch are always his intimate concerns to be left in such a dither as to the fate of something very close to him." Hopefully, "Harris Common Sense" would prevail.

Yet, when it came to the ranch, Harris common sense was in short supply. My grandfather had not discussed the future of the ranch with his children, and on this topic they hardly talked with each other. Too much emotion was in the air. Doing nothing and assuming that kindliness would prevail was the path of least resistance. The years passed. Marie, Tod's widow, remarried and moved to Rossland. Heather died suddenly of hepatitis in December 1960; her husband, Bud Rose, had died a year earlier. Effectively, the executors of Granddad's estate became my father and Sandy, neither of whom had a head for the quick, legal resolution of family conflicts. Marie would do whatever the family wished. My mother, more forward in this regard, was unwell but in the wing.

A resolution of my grandfather's will was finally reached in the summer of 1964. My parents and Sandy bought the ranch land for $4,500, a sum that, divided fifteen ways, amounted to $300 for each beneficiary. My father acquired title to the land around the cabin including the old ranch house and the slope below the old ranch road to the lake. Sandy acquired the rest.[4] The purchase price was based, loosely, on an appraisal done in 1953 of the ranch buildings and land. The value of my parents' log cabin and Sandy and Mollie's house was not included in the purchase price (although legally they were part of the ranch), and the old ranch house and farm buildings, which had some value in 1953, were judged worthless a decade later. Not a lot of money changed hands. Beyond my father's and Sandy's families (whose shares were waived), $2,400 was distributed among eight beneficiaries. I long suspected that this money came from my parents, and a letter from Dad to Sandy found in a box of Sandy's papers suggests that most of it did.[5] In the interest of reaching a settlement, Mother had offered to pay the whole sum, but my father, considering her offer too generous, suggested that he, Mother, and Sandy contribute equally. Probably they did.

Dad's letter to the beneficiaries about the offer to purchase the ranch was fulsome in praise of Sandy's care of the ranch. He began, "It [the ranch] looks like a place that has been cared for and looked after. This maintenance has all been the work, single-handed, of one of the legatees [beneficiaries], A. L. Harris." And concluded, "I think that all of the rest of us who are legatees owe a debt of gratitude to Uncle Sandy for all he has done on the ranch."

While largely accurate, this was also an attempt to assuage Sandy's feelings and a justification for transferring most of the ranch to him. After so much procrastination and family tension, my parents wanted closure, and this, it seemed, was the means thereto. Sandy was virtually given the large part of the ranch; my father obtained the remainder. To whom the "debt of gratitude" was principally owed was an open question.

With this agreement, tensions on the ranch diminished, but never disappeared. There were still those who stayed, and those who visited, and the strong feelings of the former that they did much of the work on which the pleasures of the latter depended. The roots of this feeling stretched back into ranch childhoods and experiences that now can be only glimpsed, and that the belated resolution of an awkward will could not transform. Over the years, the two sides of the family to whom the ranch had fallen would continue to use it differently and to share the land and the work unequally. In these differences, superimposed on long pasts, was an abiding tension that expressed itself most abruptly long after Sandy and Mollie were dead and Nancy was at the very end of her life.

10 SANDY'S RANCH

WHEN I WAS GROWING UP IN THE 1940S AND EARLY 1950S, MOST OF THE Bosun Ranch seemed to belong to my uncle Sandy who lived there with his wife, Mollie. It did not, of course. Until 1942, when the ranch was leased to the federal government for the duration of the war with Japan, my grandfather, J. C. Harris, had always been "the boss." His intentions carried. Although Sandy had long done much of the practical ranch work, the ranch was not his. During the last three years of World War II, and for several years thereafter, it was a federal leasehold, a camp for interned Japanese Canadians (see chapter 8). In 1949, when the government returned the ranch to the family, my grandfather was near the end of his life. Effectively, the ranch was becoming Sandy's, although legally it belonged to my grandfather and, after his death in 1951, to the fifteen children, spouses, and grandchildren identified in his will (see chapter 9).

Sandy grew up on the ranch, knew it intimately, and in 1922, when twenty years old, was sent to Ontario Agricultural College where, more than thirty years before, his father had played rugby and cricket. Sandy (who played hockey, baseball, and his violin), was expected to learn agriculture, presumably because his parents considered him the son most likely to take over the Bosun Ranch. By inclination, however, he was not a farmer. The enormous amount of work he put into the ranch had more to do with water and electrical systems, logging, cleaning up, and cutting cordwood than with farming. He worked intermittently on the ranch, but also for a time in the mines, in a concentrating mill, on a tramline, on the roads, and eventually as manager of the power plant at the falls on Carpenter Creek. His father, who owned the New Denver Water Works, gave it to Sandy, and in 1942 Sandy purchased the New Denver Light and Power Company for six thousand dollars. He had become a small businessman and owner-operator of a small company.

In 1927 he married Mollie Colgrave, and Sandy and Mollie lived on the ranch in the cottage Sandy's Scottish uncle, Charlie Raeper, built about 1902 and in which Sandy and my father attended high school (see chapter 3). Their daughter, Nancy, was born in 1930. In the fall of 1948, Nancy moved to Vancouver to attend university. As soon as she left home, Sandy began the weekly letters to his daughter that would continue for almost thirty years. Hundreds of them survive. Usually two or three typed, single-spaced pages, they are an intimate record of Sandy's thoughts and doings during the years when he managed—and then, after 1964 when my grandfather's will was finally settled, owned—most of the Bosun Ranch. Sandy and his stories are much remembered, but his weekly letters to Nancy are his principal surviving record.[1]

IN THE FALL OF 1948, WHEN THE LETTERS BEGIN, J. C. HARRIS, SANDY'S FATHER, was living with Sandy and Mollie in their house on the ranch. His eyesight was not good—new glasses had proved useless and he couldn't read—he had dizzy spells, and he was often confused. His ideas, Sandy said, were often screwy, and writing, which he often did, upset him. Sandy considered him helpless and largely dependent on Mollie (Dec. 3, 1950). There is no indication from the letters that Sandy and his father had much to talk about, or that their relationship was close.

Sandy had recently sold the water works to the village of New Denver, but was preoccupied with the New Denver Light and Power Company, supplier of power to New Denver, Silverton, Rosebery, and some of the mines. His outfit was small—he usually employed an accountant, a lineman, and a man at the plant—and supplying reliable power from an irregular mountain creek laden with tailings from the concentrating mills was always a challenge. Ice storms clogged and broke his equipment, tailings from the mills near Sandon or along the road toward Kaslo (particularly the Lucky Jim, the Cody-Reco, and the Carnegie) ruined the valves in his generators, and wind and thunderstorms played havoc with power lines and transformers. Sandy was in the thick of this—at the power plant trying to repair valves, up power poles to fix transformers—especially after his trusted employee Tak (Takayuki Hatanaka) left for Toronto in 1949, and he had to deal with unskilled and unreliable replacements: "My new lineman is turning out to be a washout ..." (May 25, 1952). After one particularly devastating storm and creek surge that shut off power for twenty-four hours in early February, Mollie appended a note to Sandy's weekly letter to Nancy. The power had been off three times that day, and she wrote beside a coal oil lamp: "We've never had so much darned trouble, poor Dad's about fed up!" (Feb. 4, 1951).

The New Denver Board of Trade had long encouraged the West Kootenay Power and Light Company, which operated a dam at the Bonnington Falls on the Kootenay River and supplied Nelson, Trail, and Rossland, to bring power to the Slocan. Sandy expected his company to be taken over, either by West Kootenay or by the British Columbia Power Commission. As it turned out, an offer to purchase came from the redoubtable Viola MacMillan, the first woman president of the Prospectors and Developers Association of Canada who, long after her Slocan adventures, would be accused of supplying misinformation about her Windfall copper–silver properties near Timmins, Ontario, then jailed for eight months for wash trading. Viola MacMillan was a tour de force. In the summer of 1952 she was promoting a Slocan mine and mill, and proposed to build a new, large dam and powerhouse at Sandy's site on Carpenter Creek to supply power to her operations and to the domestic market. Sandy suspected that she wanted control of Carpenter Creek and, by being tough about tailings, of all the mills and mines in its watershed (Aug. 18, 1952). However, after a meeting in Sandon with all the mine managers, he was reassured and decided to sell. The selling price was fifty thousand dollars for the outstanding shares of the New Denver Light and Power Company (all of which Sandy owned or controlled) and twenty-four thousand dollars for the company's bonds.[2] Sandy could manage the new plant if he wished: "They intend to put in a big dam and pipeline to the lake and do it right away. It is going to cost a lot of money and should be a good outfit. I think I would have been wrong to have held out and prevented it so my conscience is clear and I am entirely happy about it." (Aug. 24, 1952)

Viola and her husband were to pay for the New Denver Light and Power Company in instalments over a year, but shortly after their agreement with Sandy the price of base metals collapsed. Only silence came from the MacMillans; apparently they had gone to Florida. Sandy still had a power company on his hands. The tailings problem was worse than ever—"If it hadn't been for tailings, I would never have made the agreement with Viola in the first place. It's tailings that are putting the D.L.& P. on the blink more than anything else" (March 15, 1953)—and he still couldn't find reliable employees. He mused about tunnelling through a rock shoulder and building a new dam farther up the creek, but such plans would be costly: "I have looked at the gorge so much and thought so much about it that I always come back to the same old ideas. I think it needs fresh brains ..." (Sept. 27, 1953). Late in 1953, he made a deal with the Western Exploration Mining Company to jointly develop Carpenter Creek if it struck ore in a new drift. It did not. A year later Sandy was in Victoria for discussions with the director of the BC Power Commission. They agreed that Carpenter Creek was not an economic source of power (Sept. 26, 1954). The question, then, was when would the Power Commission build a line into the Slocan? Sandy waited, dealt with rumours, and was increasingly fed up: "It seems to me that all I have been doing for years is wait for some word or other from that darn Commission. It will be a tremendous relief to me to get something concrete out of that bunch. I am getting just a wee bit sick of them all. If I get much sicker I will have to go to the new Doctor" (Oct. 30, 1954).

With mixed feelings, Sandy finally sold the company to the BC Power Commission in 1956 for, probably, the price at which he had been prepared to sell to the MacMillans (roughly seventy-five thousand dollars). He had told local people they would be better off under the commission but, surveying the scale and costs of change, he reflected that "our old way of doing things was cheap, light and small in comparison to the new way ... we were able to achieve the same results about five times as cheaply as the highly unionized Govt. organization does" (Oct. 14, 1956). He hoped the public would not consider that he was trying to feather his own nest. He still considered the commission "the right way to do the job," thought that people's laziness and greed had driven up costs, and hoped that the commission idea would eventually be proved right.

For the remaining thirty years of his life, Sandy's income came principally from the funds obtained by this sale. He and Mollie lived frugally, and relied on a large kitchen garden as well as on hunting and fishing. The ranch produced some income, but never very much. Overall, a large measure of local self-sufficiency yielded enough to get by. Sandy was out in the world—president for a time of the local board of trade, a member of the school board, active in bringing television to the Slocan, a vocal critic of the Columbia River Treaty—but, like his father before him, his income was increasingly frail. He was a critic of salaried professionals who, he often thought, had little useful knowledge and lived off the avails of taxpayers. When he and Mollie became eligible for government pensions, they were not claimed. The two of them had enough; taxpayers' money could be used better elsewhere.

When the letters began in the fall of 1948, and for some years thereafter, Sandy was an enthusiastic hunter. His frequent hunting and fishing companion was Pete Nelson—Norwegian by extraction, member of a Canadian tank corps in World War II, road foreman in the upper Slocan, and a resident for years with Sandy and Mollie. Pete, an alcoholic when I knew him in the late 1940s and 1950s, probably suffered from post-traumatic stress disorder, but he could be a great deal of fun; his sense of humour was similar to Sandy's, and they shared a deprecating banter. Principally they hunted deer, mountain goat, and grouse. If bears and cougars were spotted and tracked down, they were shot, the former because of the damage they did, the latter because they were considered dangerous. Sandy poisoned coyotes, held to be killing game, with meat supplied by the Fish and Game Department. Neither he nor Pete paid any attention to fishing limits, and frequently caught sixty to seventy small rainbow or cutthroat trout at an outing in one or other of the local creeks. Fish and game were eaten or frozen.

Accounts of hunting, often laced with ranch humour, recur in Sandy's early letters:

> I SENT MY man Nelson to get me a goat today [up Ten Mile or Enterprise Creek]. I stayed home and hauled wood this morning. Nelson did not do as I ordered. He only brought home one lousy blue grouse. I imagine I will have to go with him in future if I want to get results (Oct. 18, 1948).

Pete was raising pheasants, hoping to add them to his quarry:

> WE HAVE BEEN training the pup to hunt pheasants and he is getting very good. The trouble is that we only had one pheasant for him to hunt and every time he put it up we would shoot near it to give him the idea. He was certainly catching on in great shape and it was costing us a bunch of shells. I am sad to say that that darn Nelson missed today and killed the pheasant. Now the pup's training is over (Oct. 16, 1949).

> THERE IS A coyote hanging round laughing at me. I had a shot at it on Friday. I shot from the house and he was up at the top of the clearing (where I always shoot them). I guess there must have been a hole through him because the bullet didn't seem to bother him very much. Anyhow he was round looking for me today, but before I could talk to him he had gone off to another meeting (Nov. 8, 1952).

> I WENT COUGAR hunting yesterday. I went the day before and took Pete's dog, but that wasn't serious. Yesterday I was serious and meant to get the cougar so I took your Mother. She wasn't any

better than Pete's dog and the cougar is still running loose (Feb. 14, 1954).

As the years passed, he hunted with increasing misgivings. By the early 1950s, he and Mollie were feeding deer. When a flock of geese arrived on Bosun Lake, Sandy began feeding them too, and hoped they would nest: "I haven't had anything so interesting to play with for years" (May 28, 1954). He wanted Nancy to talk with one of her professors at UBC (Ian McTaggart-Cowan) about how far coyotes range. He wasn't sure whether poisoning had been helpful, and wondered when the ranch would be restocked (March 16, 1952). When he shot a deer in the hindquarters, pursued the wounded animal but could not get a proper shot and had to leave it to die, he wrote that he would never again shoot a deer (Oct. 7, 1956), a resolution he apparently kept.

Mollie with cougar, n.d. Cougar hunts were part of the early ranch life and, as I spent summers there in the late 1940s, I remember some of them. With dogs, it was usually easy to put a cougar up a tree, where it was then shot. Harris family photo

The ranch itself was always a preoccupation, even during the years when he owned the New Denver Light and Power Company. About 1950 Sandy bought a small John Deere tractor, and loved this machine that did so quickly what had taken so long with horses or by hand. He used it to clean up the Far Field by Bosun Lake where the twenty-five Japanese Canadian houses had been during the war. The government had demolished most of the buildings, but left rubbish, root cellars, privy pits, fence posts, and ridges between former gardens. With his tractor always at gentle throttle, Sandy cleaned up this mess. It took more time than he had, but he enjoyed himself. He told Nancy that she "would have fun over there with me … I have a white coyote for a pal but he isn't much help … The coyote just sits and watches and sometimes he sits and laughs. One of these days I will put a bullet through his vitals if he gets too insulting. He is the lightest colored coyote I have ever seen" (Oct. 16, 1949).

He also enjoyed spring cleanup and burning, especially with his tractor:

> [W]E HAVE ALL been … busy clearing up and burning. You would have had a whale of a time today and yesterday. We have done a lot of clearing round the old house. Fences are taken down and burned, the old blacksmith shop is gone, and the little house by the power house is cleared away and the hole filled … The creek bed from the barn to the cathedral [the carriage shed near the old house] was full of grass and roots so that the creek wasn't running in it at all … I used the bulldozer and cleaned it all up, pushing stumps out that have been there since Granddad was a boy [that is, a young man] … Gosh I like clearing up after winter. The trouble with me is that if I was a farmer I would spend all my time clearing and not producing anything … (April 5, 1952).

Other ranch jobs were less pleasant but had to be done—shovelling snow off the barn roof, for example, to keep it from collapsing. He wished he had more time, especially to work with his Cat: "I love to clear up with the cat, the trouble is that it doesn't pay wages and costs money" (April 5, 1953).

Always in the back of his mind—and more to the fore after the sale of the power company—was the question of what to do with the ranch. Sandy had lived through his father's many attempts to create a viable commercial farm, and had concluded that none of them could work. The ranch was too small and isolated for the economies of scale associated with modern farming and marketing. What, then, was it good for?

The earliest answer in his letters was that the ranch might become a deer farm. Most of it was too rough for cattle or sheep, but could support a great many deer ("wild cattle" he sometimes called them) if they were fed through the worst of winter. There would be problems with predators and poachers, but he asked Nancy to talk with Dr. Cowan about deer farming (Jan. 3, 1952). In one respect, Sandy's plan worked exactly as he thought it would. Supported by winter feeding, the deer population on the ranch exploded. But it was illegal to raise and harvest deer; moreover, Sandy became so fond of his deer that harvesting them was out of the question.

Christmas trees were another option. The trick, Sandy thought, was to find the right trees. Douglas firs grew too fast, their branches were too far apart, and their needles were not green enough. He needed more information and doubted he could get it from the Department of Forestry: "[T]hey are all Government employees and get salaries. People who live on salaries are not apt to have good productive ideas because their lives don't demand it of them and they deteriorate ... I want to take a trip to E[ast] Kootenay just to investigate the Xmas tree business and see for myself how it works" (March 29, 1953). Because subsequent letters make no mention of Christmas trees, the results cannot have been encouraging.

Sandy liked logging, and toward the end of his life often said that the ranch was only good for growing trees. When he acquired the Caterpillar tractor, he began building hauling roads into the forests above the ranch fields. Logging trucks often came to pick up two or three loads, sometimes several more. But a little logging on the ranch was one thing, logging there as a vocation quite another. A lot of the trees, Sandy said, appeared healthy, but on cutting proved diseased and worthless: "The trouble is that a fellow spends more time making roads the trucks can get in on than he does logging. There just aren't enough good logs to make it really worthwhile" (March 30, 1958). He also thought that poor early logging had not prepared the ranch for sustained-yield forestry, but hoped that when the roads were built "it shouldn't be too hard to log the whole area in a decent manner that will allow the timber to grow almost as fast as I want to take it off" (Oct. 3, 1957). Basically, he wanted a steady, annual crop of logs. However, the ranch could not provide a living from

Abandoned hay barn and parts of Bosun Lake and the Far Field, August 1972. The water in the foreground is part of the enlarged lake Sandy made with his tractor. The barn, a ruin in 1972, had been built years before at an edge of the Far Field where, during the war, there was a camp for Japanese Canadians. Harris family photo

logging. Some thirty of its approximately three hundred acres were in fields, a good third of the rest was too steep to log, and the timber on the remainder was spotty. At one point Sandy investigated the possibility of acquiring a tree farm large enough to create "an interesting permanent job" (Oct. 26, 1956) in a proposed forest reserve between Carpenter Creek (at New Denver) and Four Mile Creek (at Silverton), but nothing came of it.

Then there was the Bosun Lake. In the mid-1950s Sandy put eastern brook trout fingerlings in it, and a few years later rainbows. They thrived, growing so rapidly that their length hardly kept pace with their girth. The lake, Sandy realized after examining the contents of fish stomachs, was full of freshwater shrimp and daphnia (water fleas). This discovery provoked years of experimentation about when to stock with fish of what species and size, when and how to fertilize to increase the number of shrimp and daphnia, and when to harvest. For the most part, he thought of these fish as food for the family, but the larger prospect was tantalizing:

> I AM CONVINCED that the right way to utilize the Bosun L. is to buy fish about 4″ in length, in lots of about 200 about twice a year. This would ensure a goodly supply & the lake should be fished as hard as we [the family] have time for. The number bought and taken could be regulated by experience. It is my opinion that the lake is ready for about 200 4″ rainbows this fall. This thinking is based on the fact that the fish are not being fed. If they were fed, the number could be increased into the thousands (Sept. 15, 1968).

He even considered selling trout spawn: "I begin to think that the Bosun L. might be useable as a place to raise Rainbow Trout for their spawn. I have an idea that it could be made into a real dinger of a business. One would have to create a market in the U.S. I think so I'll look further into it" (May 17, 1964). Yet he never sold fish or spawn.

And what of the ranch fields? Sandy spent a lot of time tidying them up and once planted oats on much of the Far Field, and clover on the bottomland near the ranch house. But bracken sprouted among the oats, burdock among the clover, and both had to be cut back by hand. Such plantings generated work, not income. Sandy came to the view that the ranch fields' only possible commercial crop was hay. He thought that, properly irrigated and fertilized, they could produce a lot of hay, and in the spring of 1962 was delighted to sell the whole crop in advance for one-third of its delivered value. The buyer supplied the machinery, harvested and hauled the hay, and paid cash. "The payment," Sandy wrote, "was based on the ability of the ranch to produce 100 tons a year. About 3 tons per acre. With irrigation and fertilizer, I'm sure it can do this and probably more. Whatever comes of this I'm happier about it than I have been about anything connected with the ranch for years. At least the place is going to get a chance to be properly used for once, & I may

Baled hay in a ranch field, August 1964. Hay was a possible Bosun Ranch crop, but the market for hay was uncertain, there was not enough equipment to irrigate properly, and the cost of modern mowers and balers was more than a small farm could bear. Harris family photo

get a chance to go fishing" (March 10, 1962). Yet with the equipment he had—two-inch aluminum pipe and sprinklers—and a lot of work Sandy could not irrigate properly more than an acre a day, an inadequate rate for almost forty acres of hay fields. Moreover, the agreement to purchase did not last, and Sandy was soon casting around for buyers: "It begins to look as if I can't get the hay cut this year. I think that mostly the trouble is that logging has furnished people with such high paying jobs they can't be bothered with anything as ordinary & simple as hay. Woe is me ... There is a bumper crop too" (July 10, 1966). Hay came to seem yet another unreliable ranch crop.

A few years later and after prompting from his son-in-law, Sandy was reconsidering the possibility of beef cattle: "I'm glad to hear you [John Anderson] talking of beef cattle ... Beef is the only thing I can see for this place. Probably have to get a D6 and log and clear the flat above this & make Summer pasture of it. It should be left in timber but big business has put the small timber man out and I think beef is the only alternative.

I wish I was 40 years younger" (undated but probably 1969). In the mid-1970s, Sandy's son-in-law began pasturing beef cattle on the Far Field.

None of these schemes solved the ranch's underlying economic problems, which proved as intractable as they had for Sandy's father. The ranch produced a little income now and then, but was not, and never had been, a commercial success. Yet for someone like Sandy—however much he grumbled that there were too many people round, particularly the incompetent Harrises—its yield was not so much measured by the market as by the more intangible pleasures of living within a diverse local economy in a remarkable place. Such thoughts were on his mind when, in the fall of 1953, he returned from a trip to the Peace River region:

> WE WERE GLAD to get home again. New country is interesting to see, but home looks pretty good to me. I would much rather live in a place like this where you can furnish nearly all the things necessary to live on, and get a whale of a kick out of doing it, than live in a place where you produce only one commodity and buy everything else. To me, producing what you want for yourself is more like a game. Growing wheat or cattle is more a straight business. This suits me nicely thanks. There are enough interesting things to do on this place to last one person six lifetimes if he likes it, and I like it (Oct. 8, 1953).

Some years later he wrote in a similar vein to his daughter and son-in-law:

> [T]HE RANCH IS the most valuable and sensible asset that the A.L. part of the family has, and we should be careful not to get ourselves involved to the point where we could lose it. The value of the Ranch is something that money doesn't enter into. The Ranch is a way of life & it's very good when one looks around at the other ways of life and sees how much unrest and uncertainty there is (Dec. 6, 1967).

This was Sandy's basic conclusion about the ranch: it was a way of life.

Although Sandy was president for a time of the local board of trade, a member for years of the New Denver School Board, and active in the TV Society (he installed some of the mountainside discs that gave the valley television reception), and although he and Mollie curled and played a lot of golf on the New Denver course that he had helped to build, it is clear from his letters that he and Mollie wanted to be left alone far more than they were. The problem was that Sandy was engaging and charming, and people kept dropping in: "Mums and I started to have a quiet day all to ourselves yesterday, but it was not to be. There must have been nearly twenty visitors here from noon on. I wish I had my cabin in the hills where no one could find me" (May 2, 1952). He could be very critical of me and my university friends when I brought them to the ranch, as well as of the numbers at the cabin. As the years went by, the desire to be left alone expanded into a critique of the ways of a modern, bureaucratizing world and even of people in general whom, Sandy concluded, were usually motivated by greed.

Whereas his father admired the ordinary worker, whom he considered the backbone of any society, Sandy admired small businessmen. He thought that, forced to rely on their own ingenuity, they were the creative, practical thinkers in a society, yet were being squeezed by a public bureaucracy supported by excessive taxation:

> THE BIG THING against the private operator these days is that most people are working for big concerns or the Govt, & the natural consequence is that the private chap is taxed more & more, because somebody has to pay for those public earners. I have watched this building up to a point where the private punk is losing all his advantages that he makes for himself through hard work & thinking, to a state that makes him think it's probably better not to be a private punk. I don't know what is right for most people, but I hate to see the private operator go. I know that I have a lot of

> respect for anyone who can make any kind of a business pay well enough to equal the average professional income. It takes a pretty smart man to do that (Oct. 6, 1957).

He thought that salaried people were not original thinkers, and that schools and universities were turning out graduates with little or no practical experience. When his daughter announced that she planned to become a teacher, Sandy's response was ungracious. "Its a grand and noble profession second to none and the government will always see to it that the poor down trodden taxpayer will have your monthly stipend ready each month. Someday I would like to hear of someone who graduated from university and who still knows enough to be able to earn a living on their own hook" (July 20, 1952). When on the school board, he usually opposed teachers' salary demands.

In many ways, Sandy's politics were those of a small businessman. He was not a political visionary, and could be sharply critical of his father and his more radical associates (see chapter 6): "I remember the days when old Woodsworth, Doc. Telford, Angus and Grace MacInnis, and Coldwell used to arrive at the Ranch for a policy meeting on the old CCF.[3] I think that, all told, there weren't more than 20 people in all of Canada who were really interested in what these few guys were trying to put over" (April 25, 1971). On the other hand, he voted federally for Bert Herridge (Kootenay West) and provincially for Bo Harding (Kaslo–Slocan), both old CCFers, and had little use for the province's Social Credit government. He told his daughter that he was "an individualist who thinks that socialization is the right thing in many cases," a combination that "is bound to have one in a mental mess" (July 20, 1952). And, like his father, he thought that something was deeply wrong with the world, "the sort of thing that Granddad Harris spent his life trying to understand. It's the main thing that makes Muz and me want to stay up here ... It's why we like plants & animals better than people" (Dec. 6, 1967). He thought that plans for social change would fail because most people were inconsistent and greedy—because human nature itself stood in the way. From time to time he declared himself completely "fed up with people. If you want to know what I mean just listen to the World News a few times. It's all there & in every direction" (Oct. 31, 1970). Rather, he would "stick with Nature. She was tough but consistent." That said, late in his life he spoke admiringly of his father's political work, told stories about the intersection of the ranch and the CCF, and encouraged his young hippie friends to support the New Democratic Party.

Yet, increasingly, Sandy did "stick with nature." He had always been interested in the creatures of the ranch, but from about the mid-1960s his focus turned more and more to the natural world around him. He observed nature closely, and loved and was fascinated by wild animals, none of which—except possibly bears when they got into his fruit trees and muskrats when they interfered with the ducks and geese on Bosun Lake—he any longer had any interest in shooting. He fed deer, fish, coyotes, squirrels, and grouse, and named and got to know many of the animals individually. Increasingly they relied on and trusted him. The ranch became an un-caged menagerie stocked by the Slocan's more accessible creatures. Sandy mused that its ultimate use might be as a game reserve. He thought the Game Department was beginning to realize that "there are probably more game animals right here than any other place of comparable size" (March 2, 1969).

Deer are among the principal characters in his letters from the late 1960s and early 1970s. Blacktails (mule deer) came to his lawn and even to his door to be fed, and did so winter after winter. The whitetails were more skittish, but Sandy was working on them. It was an event when the deer came back: "On Friday night, just after supper, I looked on the front porch to see if Chippie Hackie & wife [flying squirrels] were there. I opened right on Old Nellie coming back this year for something to eat off the porch. Gosh we were pleased to see her ... Nellie cleaned up all the flying squirrels' food and b-----d off. I put food out for her last night but she didn't come round. I haven't any apples or crushed oats. I'll have to get a sack of oats. We have to give our friends the best" (Dec. 17, 1967). Such treatment had the desired effect: "Our deer friends are getting more and more friendly. They line up and take food from my hand in dignified

Sandy and his deer friends, ca. 1975. Sandy at the back of his house with some of the members of what he described in a letter to his daughter as the "Board of Wild Life Management." He named and individually recognized most of the deer on the ranch. They were pets. Pond family photo

dignification. I can put a cabbage leaf in my mouth and Old Grizzly [an old blacktail doe, one of Sandy's favourites] will step up ... and take it away from me" (Jan. 17, 1971). This was all very well, but Kluane (Klu, Klukie), Sandy's Tahltan bear dog, was less sure about Sandy's deer friends: "I have been working on Klukie and the deer. I think Klukie is a bit jealous of them & she was monkeying around while I was feeding. She got a bit too far away from me & Old Grizzly decided to teach her a lesson. She made a real pass at Klu, who ran behind me. I was on the front porch. Klu said 'by G-d Unc. The old B---h is really hostile.' Actually Klu & our tame ones are getting on very well ... I think they sort of like each other. They respect each other both ways" (Feb. 14, 1971).

When Old Grizzly died three months later, after being chased by dogs, Sandy reported how much he and Mollie missed "the old girl," dependent as she had come to be on them for food (April 3, 1971). With Grizzly gone, what Sandy called "The Board of Wild Life Management" that met on the lawn and now had lost its matriarch consisted of Peanut,

Popcorn, Flatback, Lady, and Mae West, all female blacktails. These deer, and a slightly larger number of whitetails, had become his welcome responsibility. This is late February 1972:

> IT IS A very wet morning. I got up & fed 18 deer twice ... I see they are all back looking for more now. Disraeli is here. He still has his horns on. Gladstone is here a lot now too [Disraeli and Gladstone were the patriarchs of the blacktails]. We have three little runts. I surely like them & we have fun fooling the old ones so the runts get most food ... Peanut is always there and gets the best of everything. You can pick Peanut out of the herd at a glance because she is in so much better condition than any of the others. The gang is getting roughly $10 worth of concentrated food a week. I don't know if it's enough or not. They don't think it is. They also get a lot of cedar and about 100 lbs of alfalfa a week. I have trained them to eat range pellets for beef cattle (Feb. 20, 1972).

One of the reasons Sandy fed deer on his lawn was that his wife Mollie's memory was failing. She could spend hours losing and finding the same thing, could forget where she was, and was beginning to wander: "Last week I suggested that she wash the kitchen floor with the mop. She got going and darned if she didn't do the bathroom and den as well as the kitchen & she did them well too. She was very pleased with herself and got hugged and kissed all to bits. Now I have to try to get her clothes ready for tomorrow. Gosh, she can't remember where anything is and I have to be very careful about asking her help or I make her feel helpless and that makes her depressed" (Oct. 21, 1972). The housework, and increasingly the full-time care of his wife, had become Sandy's. At times it was difficult to get out of the house, even to the garden. In these circumstances, the deer, usually at hand, always intriguing, and requiring no act of memory, were indispensable: "Muz gets more pleasure out of the deer than anything else and they are very valuable to us" (March 18, 1973).

In the fall of 1969, Sandy was feeding Selkirk, a coyote whose mate had been shot. Perhaps Selkirk and Kluane would get together, and she would have a litter of part coyote pups (Nov. 16, 1969). Sandy was hopeful, but the relationship was not consummated. Among the coyotes around two years later was an "absolutely beautiful" young coyote: "I'd surely like to have a coyote like this little pup. I've heard so much guck about how untrustworthy a coyote is. I'd like to find out for myself. I'll bet that one raised from a very young pup would be just as safe and good as any dog" (Sept. 6, 1971). A year later there was Aurore, as beautiful, Sandy thought, as an animal can be. She "came for food every night last week except the last night. Probably someone shot a deer on Saturday and she has all she can eat. I hope they don't shoot her. I'm going to make a contract with her to supply me with coyote pups. She has been eating exactly the same amount of food that Kluane gets ... Gosh she is a thing of beauty in the sunshine with her new winter coat on. We were in the jeep and Klukie bawled her out. She just sat down and watched us from about 20 yards." Sandy was enthralled by Aurore: "I don't know how many times a day I look to see if she is sticking her head over something to watch me and Klukie. She is on Klukie's mind all the time too. As soon as she is turned out, she looks to see if Aurore is there" (Dec. 2, 1972). Selkirk and Aurore were imagined pets.

And there was Chunky, the ruffed grouse that adopted Sandy. When Sandy was logging or brushing in the woods Chunky would appear, run around, and talk for hours: "The darn little fool will stand talking with his head as close as 4″ to where my axe is making contact to trees & when it strikes right by him, he doesn't even blink. He has a lot more faith in me as an axe man than I have myself ... He's a great little pal and he follows every move I make, running forth and back every 10′ all day. I find he isn't as happy when someone is with me and he doesn't like people talking loud" (Oct. 30, 1960). Chunky would eat from Sandy's hand, ride in the cab of Sandy's truck, was briefly in his house, and was probably the most forthcoming of all his wild creatures:

Sandy feeding Chunky, n.d. The ruffed grouse seemed to have adopted Sandy, spent hours chatting with him when he was working in the bush, and would try to draw him under a log whenever a hawk was around. Harris family photo

I MUST SAY that through knowing Chunky I am able to learn a great deal about grouse etiquette that I never knew before & I doubt if many other people are aware of just how keenly alive a grouse's mind is to what is going on in the woods around it. I believe that Chunky knows of everything extraordinary that is taking place within 200 yards of her [Sandy variously identified Chunky's gender], & if she doesn't understand it, she will investigate till she does. I think she is about to build her nest, and I am going to try to locate it so I can keep an eye on proceedings (April 24, 1960).

From time to time Chunky would try to herd Sandy under a log, and he eventually realized that whenever she tried to do so a hawk would soon pass overhead.

Sandy also fed and somewhat tamed Hondo, a red squirrel; Chippie Hackie and his mate, a pair of shy flying squirrels; occasionally Old Pork, a porcupine; and the geese and ducks on Bosun Lake. He enjoyed them all, but not more than the fish and other creatures living in Bosun Lake. The Bosun was his lake. He had doubled its size with his Caterpillar, turning a grassy marsh between the original lake and the Far Field into a shallow extension of the lake, and spent as much time there as he could. He would lie on a raft for hours, peering into and absorbed by the contents of a lake that was "teeming with billions of minute forms of life that only show up in the sunshine & when something the right color is behind them. What a place that lake is!!!" (Jan. 10, 1965).

Sandy loved his lake, which seemed for a time to have commercial possibilities, but was a fascinating biological realm in its own right. The biological challenge that most interested him was to encourage the trout he had introduced to spawn, and thereby make the trout-populated lake a self-perpetuating biological system. To this end he created what he called his "maternity ward." It was a long, narrow bed of gravel with a hose at one end to create a current. Sandy worked hard on and was proud of this maternity ward, and reported on it effusively to his daughter:

> MY TROUT ARE, or were, spawning. I have spent a great deal of effort to build them the kind of beds their little hearts desire. No human ever had more loving care and forethought than they have had. They have every kind of bed any trout could ever think of. I have been all over the country ... & I have conscientiously sought out and delivered to the maternity ward every kind of bed from soft mud to ¼″ crushed quartz which looks like snow and should warm any darn female fish's heart. They also have material up to 1″ which has been polished to a beautiful glow by Mother Nature (May 10, 1964).

The maternity ward worked; trout spawned there, but Sandy found that leeches ate their spawn. He began to trap them and in any given year filled several gallon buckets with leeches, but still they came. "It is no wonder," he wrote ruefully some years later, "that very little reproduction is able to take place" (June 16, 1969). His other overriding problem was that he could not control the lake level, and consequently his maternity ward might be high and dry or fairly deeply submerged. When enlarging the lake, he had extended it over some of the old workings of the Bosun Mine, and the lake may have been draining through an abandoned shaft. Sandy tried for years to fix this, never successfully. The lake he had partly made was fascinating, but hardly simple, and as Mollie's memory diminished in the early 1970s and her care became his preoccupation, he had less and less time for it.

IN THE LATE 1960S AND EARLY 1970S, AMERICAN HIPPIES WERE POURING INTO the Slocan. With them came a strong reaction against what they judged a materialistic and often corrupt American culture and a desire to live simply and close to nature. Most of their backgrounds, however, were urban, and they came, usually with very little money, into what for them was a radically strange environment. New skills had to be learned and difficult adjustments made. It was against this background that some of them encountered Sandy. He was not of their generation, had neither long hair nor a beard, had never taken drugs, and his devotion to and care of a wife who could no longer be a companion was close to the opposite of their common gender relations. But he was completely of the place to which they had come. He knew its stories, its land, its animals. He knew how to move around and live in the bush, how to create and maintain a garden, how to log, how to repair equipment. He knew where things were. Moreover, he was kind, and he enjoyed, and often helped, most of the hippies who crossed his path. For their part, Sandy, whose whole life experience was so different from theirs, seemed to embody much of what they themselves hoped to be.

Sandy particularly enjoyed three Californians, Peter Pal and Ruth Vitner, whose wedding he witnessed in September 1971, and Matthew Hudson, who often came to help with the hay. However, his closest connections with the hippies came through the ranch house and those employed to work, and encouraged to live, there (see chapter 11). Thomas Wright, originally from North Carolina, had worked on the house through the summer of 1972, and after a winter in Sandon returned the following spring with his wife, Lilly, and baby, Autumn Maple. They tented below the ranch house beside Slocan Lake, but wanted a garden close to the old house and to live there the next winter. Sandy's advice was needed: "Thomas Wright is working hard on the ranch garden. He has a terrible job on his hands. It should have been worked over last fall. His wife gets out there with him & they work like blazes. I don't think it's very well directed work but he tries. It's the effort of a city man trying to be a country bumpkin all at once. He keeps asking me questions, & I try to help him, but I think he has to learn from his own mistakes. He can laugh at himself and so there is hope. I think we like them both and their little girl ... is a cute happy little runt that likes to eat dirt" (May 12, 1973).

Lilly remembers this differently.[4] She and Thomas were attacking the garden when Sandy arrived on his tractor. "So you want a garden?" "Yes." Three days later Sandy had ploughed a large garden and suggested they roll back sod along the edges to make borders and plant squash. He would come over to talk about compost, chicken manure, and fish fertilizer (in this case fish manure added to water). Sometimes he brought algae and weed, obtained from Bosun Lake by a linked chain drag pulled by his tractor. He and Thomas had long conversations about water, which he said the R. C. Harrises did not begin to understand. He advised on what to plant and where, and, later, on how to prepare a garden for winter. Sometimes he brought lemonade. Lilly said that he had a "gift of gardening," that he was their teacher, that "he showed us how." He taught her, she said, how to do permaculture. Once Sandy asked them what they would do if cities collapsed and people came looking for food, to which Lilly replied that she would keep her knowledge, not her food, a reply that pleased Sandy. When Mollie was lost, as frequently she was, he came over for help.

The old house was hardly habitable in the winter of 1973–74. In early November, Thomas went to Calgary to buy an assortment of double-glazed windows, which he proposed to sell, and the temperature suddenly dropped: "He left Lilly and the peanut alone in the ranch house with only plastic windows and the cold snap caught them ... I had asked her to stay here but she didn't want to be a bother. You have to hand it to those people for trying to do things that others are too well off to try" (Nov. 15, 1973). Sandy was greatly relieved when Thomas returned and, soon after, Lilly and Autumn Maple left for California. She was back in the spring to help Thomas with the garden, but their marriage was breaking up, and they both left at the end of the summer. Lilly remembers the ranch as a spiritual place, a sanctuary. Sandy was a gentleman who taught her how to garden and greatly influenced the course of her future life. The deer and other animals were his friends.

When Gretchen Jordan, an artist and weaver from Pennsylvania who had come to the Slocan in 1971, and her two young boys, Aaron and Ischi, moved into the ranch house in the fall of 1974, there were two wood stoves and glass windows on the main floor, but many cracks around windows and doors and still no main chimney and fireplace or window glass upstairs. Much of Gretchen's winter went into the struggle to keep warm.[5] Sandy ploughed the road for her, split some of her wood, and came over to chat as often as he could. He seemed endlessly knowledgeable about practical matters like selective logging, and full of stories about the early days of Sandon, the ranch, and Granddad and the CCF. He could talk for hours given half a chance. He made her feel safe. She was never fearful, and never felt alone, even when there were cougar tracks in fresh snow around the house. Sandy was at hand, and was keeping an eye on her. And she felt that she gave something back, helping to look for Mollie when she went walkabout, having Mollie over so Sandy could get at his chores, and consoling him when, grief-stricken, he finally took her to a home in Vernon because he could no longer look after her.[6] Cold as the old house often was, Gretchen loved the ranch—beauty, she said, was her food—and considered that Sandy, so "kind and generous of spirit" and so at home with land and animals, represented the very ideals of the counterculture.

Gretchen and her boys left the ranch in June 1975; some six months later, Nancy and her husband, John Anderson, moved in with Sandy. For years they had planned to build their own place on a promontory well above Sandy's, but with Mollie gone and Sandy living alone, they decided to live together. The three of them shared Sandy's house for the next ten years. Even when Sandy had a pacemaker, he went out each morning "to read the newspaper" (to see what had happened the night before on the ranch) and to feed his deer and fish.[7] Then in late December 1985, he

Sandy and friend on the John Deere, early 1980s. This machine, Sandy's second John Deere, was much larger than his first. He ran them gently, at very low throttle, and accomplished a great deal of work. His friend was my family's cairn terrier, another of the many animals that were completely comfortable with him. Harris family photo

tipped over in a snowmobile and was badly hurt. He was sent to the UBC Hospital in Vancouver, where complications developed and he died on January 26, 1986. He was eighty-three years old. Gretchen was driving in Vancouver when her cellphone, an early model about the size of a brick, rang. A friend was phoning with the news that Sandy had died. She pulled over, turned off the engine, and cried her eyes out.

AS I LOOK BACK ON SANDY'S LIFE, I AM STRUCK BY ITS DETACHMENT FROM THE England and Scotland of his parents. The stuffed, brocaded comfort of Harris life in Calne, Wiltshire, was not at hand, nor even his mother's far more sparse artisanal world in the village of Fyvie, near Aberdeen, Scotland. In Sandy's childhood, there had been attempts to hold on to old ways. Care packages of English children's clothing came out from Calne;

for a time Sandy and his elder brother headed off in starched white collars down a snowy path to school. Such transplanted ways had not held. The societies that surrounded life in Calne and Fyvie were not at hand, nor any much like them. Most people in the Slocan Valley had British roots, but Britishness had been jumbled there, often after several generations tinged with Protestant Irishness in eastern Canada. Nor was the social hierarchy as defined or steep as in the societies his parents had left. The landed elite were absent, and in a valley where the economy sputtered as the mines closed, wealth was not being produced and most people were doing little more than getting by. Ethnicity and race, rather than class, tended to be the primary bases of exclusion. The land itself had no close British equivalents.

Sandy's life is inseparable from its physical setting, which was virtually the opposite of the deeply known, ordered, and managed land his parents had left behind. He had grown up on a raw farm recently carved out of a forest on a narrow, proglacial terrace on a Kootenay mountainside—a place that not long before Sandy was born no white person had ever seen. He roamed the new land at hand, knew it as well as anyone in the modern Slocan, and reflected its opportunities and limitations. Embedded as he was in a different land and society, his life could not come out English or Scottish. Being in a different place has consequences.

I am struck too by how deeply he was part of the Bosun Ranch. Except for a boyhood year at St. Michaels University School in Victoria and two years at Ontario Agricultural College, he lived there all his life. In a new society recomposing itself at a margin of empire where places were being created and abandoned and flux was commonly at hand, Sandy's life was unusually rooted. There was no opportunity in immigrant British Columbia to live, in situ, with a long past stretching back through the generations, as so often there had been, and in many places still was, throughout western Europe, but Sandy had pushed the connection with his particular place almost as far as his immigrant society allowed. The ghosts and leprechauns that the English poet Rupert Brooke missed when he sat above Lake Louise in the Canadian Rockies and reflected on what it was to live in a new land were not at hand,[8] but Sandy's world was teeming with the accumulated experience of a situated life. He lived within that experience, particularly as it expressed itself in his stories and lore. More and more as the years went by, he became a storyteller who loved his favourite stories, and repeated them with wit and relish. During the last years of his life, most of his stories concerned his deer, his coyote friends, Chunky his pet grouse, and the fish in Bosun Lake.

For all the efforts to the contrary, the ranch, it had turned out, was not a viable economy, and Sandy's relationship with it became essentially playful. He loved and played with Bosun Lake and its fish, and with the animals he somewhat tamed. He played, often very usefully, with his Caterpillar tractor. Cleaning up after winter and even after the Japanese Canadians was fun. Spring burning was fun. The geese that arrived on Bosun Lake were fun. Sandy's letters are explicit: the ranch was not a business, not a monoculture, rather an enormously varied playfield that principally yielded food from a kitchen garden and pleasure. Even in his father's day, when a great deal of hard work had been devoted to the ranch, fun had always been in the air. Nor had there been any sense, common on pioneer farms across the span of Canada, of a life and death struggle to tame the wilderness and create a farm that could support a family. Behind Sandy's father was English money, and although his father worked exceedingly hard, and although over the years the money diminished and almost disappeared, it was a buffer that opened space for fun. Sandy had far less backing, but when he sold the New Denver Light and Power Company for seventy-five thousand dollars, he came into money that also diminished while also creating room for fun. If there was an English upper-middle-class inheritance in Sandy, it was perhaps this, an enhanced capacity to play and a relative indifference to money because there had been enough.

If play is what he did, he did it superbly. He opened the ranch to a type of exploration it had not seen before, and drew others into the excitement of his discoveries. He found endless pleasure in the local world around him, and when he died one of its basic ingredients was gone.

11 THE COUNTER-CULTURE AND THE RANCH HOUSE

AFTER MR. TANAKA, THE LAST OF THE RANCH HOUSE'S ELDERLY JAPANESE Canadian men, died in December 1951, the house was no longer lived in and quickly deteriorated. The shed at the back collapsed, walls began to sag, the roof leaked, and the house became a sprawling ruin. From time to time, the family considered burning it down. A match to an oily rag on a winter night would do it, but my father tended to end such discussions with "well, perhaps," which meant not for the time being. So the ranch house sat and declined. There was never a question of restoration. It was too big, too impractical, too costly to renovate. And for what purpose its eighteen rooms? The alternatives to burning it down were to let it rot and collapse or take it down. In the summer of 1969, my father and I, and before long my cousin John Rose, began its demolition.

Work had not proceeded far before the log house my grandfather built in 1897 emerged from under the framing that had enveloped it for more than sixty years. Moreover the cedar logs, long protected from the weather, were in excellent condition. This discovery transformed our options. I felt that my grandfather's original house, rediscovered and apparently sound, could not be demolished, and that we should use wood salvaged from the frame additions to restore it. Muriel was willing but, given the stench of pack rats, dubious that the house could ever be inhabited. In late August, we had to leave for Toronto, where I taught at the university; my father and John Rose continued the demolition until a large part of the frame house was removed. Most of the rest went the next summer. We saved the ornamental brackets, and most of the

Demolishing the ranch house, summer 1969. The frame house had been well built on uncertain foundations. The log and frame parts were pulling away from each other. Harris family photo

interior panelling, the interior and exterior siding, and the two-by-fours, but threw an enormous amount of splintered wood on the burning pile.

Left behind was a decrepit log building on a crumbling foundation. Nothing was square, and the walls had been cut open in odd places to accommodate additions. There was a small creek in the basement. The log work itself was poor: gaps between some of the logs were two or three inches wide, the floor joists for the upper floor were on six-foot centres. The house had been built in a few weeks in the fall of 1897, in good part by a young Englishman who had not grown up with an axe in hand. Inside this log shell were the remains of an Edwardian house: pressed tin ceilings with pack rat nests above; plank walls covered with muslin and wallpaper with pack rat nests in the cavities between planks and logs; a

The ranch house as a ruin, summer 1966. By this time the ranch house had been uninhabited for fifteen years. Snow had collapsed parts of the roof; inside, pack rats had taken over. Harris family photo

narrow, enclosed stair to the second floor; a main floor divided into two rooms (once living and dining rooms); a second floor divided into three. In places the tongue-and-groove fir flooring was almost worn through.

In the early 1970s, these unpromising remains converged with the American counterculture. Young Americans at odds with the Vietnam War and American materialism were pouring into the Slocan, drawn by a beautiful valley, inexpensive land, and abandoned but habitable buildings. Previously, I had been absorbed as a young faculty member at the University of Toronto, and Muriel by a Ph.D. in genetics and two children. The tumultuous sixties had largely passed us by, but caught up with us belatedly in the Slocan. Although we never became part of hippie culture, our friendships with many of the hippies were close, and our respect for their talents real. As it turned out, the restoration of the log structure around which my grandfather had wrapped an Edwardian gingerbread house depended on them. I initiated the project, but they did most of the work, did it well, and left a building that reflected their time in the Slocan as much as my family's values and tastes.

The arrival of the hippies coincided with our growing need for a place apart from the cabin beside Loch Colin. The opportunity to return to UBC had come, and in the summer of 1971 we returned to British Columbia for keeps. My mother had died of heart failure four years before, leaving my father shattered and struggling to live alone. Moreover, Muriel's

The remains of the ranch house, summer 1970. This is the ruin that we undertook to restore. Nothing was square. There was a small stream in the basement. Thomas Wright said the house was held up by memory. Harris family photo

parents were in Vancouver, and she was an only child. There were good reasons to return home. In the event, my father remarried in the fall of 1971 and when the following summer Muriel and I, our two boys, and Dad and Marjory converged on the cabin it was immediately clear that coexistence there would be emotionally fraught, particularly for me and Marjory. I had loved the cabin for its beauty and as the hearth of our family. Moreover it was soaked in memories of my mother. For me it was sacred space; changing it seemed like painting over a Rembrandt. Marjory, willing to spend a good part of the year in the cabin because her new husband wanted to be there, sought to make it comfortable and attractive in her own, far more urban terms, a reasonable demand that my father was willing to accommodate. The cabin began to change, and to be lived in as never before. The summer of 1972 was tense, and the project of restoring the log core of the ranch house acquired momentum.

It was in these circumstances that my father, hearing of a young carpenter living in Sandon, suggested we look him up. We did, and in the middle of the road just beyond the derelict town found a bearded, somewhat dishevelled fellow squaring a tamarack log with an adze. This was the Thomas Wright (Arrow, as he preferred to be called) mentioned in chapter 10. He listened to our account of an old log house in need of restoration, agreed to come and have a look at the place, and did a day or two later. Thomas had bearing and presence, seemed knowledgeable, and we liked him. He declared himself intrigued by the ruin before him, and agreed to undertake its renovation.

Later I learned that Thomas was born in Old Trap, North Carolina, in 1942, the son of a Methodist minister, and grew up in the state.[1] After a year in college, he joined the navy to avoid the army draft, and spent four years there (1962–66). In 1969 he was in Los Gatos, California, supporting himself as a woodworker. Among other things, he made wooden flutes, and one day peddled several of them to a small gallery and wine shop. Its owner, Lillian, bought them, was attracted to this stranger with long hair and magnetic blue eyes, and told a friend when he left that she could marry him. They were soon together. Among their friends was a sculptor who had just bought forty acres for four thousand dollars from Doukhobors in the lower Slocan Valley near Winlaw; he encouraged them to go up and have a look around. Lillian sold her little shop, and with six thousand dollars and an old truck, they headed north.

Deciding not to stay with what Thomas called "the Winlaw crowd" (according to Lilly, Winlaw felt too much like California), they explored north along Kootenay Lake into the Lardeau area, and then, driving west from Kaslo, encountered Carpenter Creek (its name, Lilly said, convinced them to stay[2]), the largely abandoned former mining town of Sandon, and its few inhabitants. One of them, Bill Barlee, who had grand plans for Sandon's restoration, owned the former train station, then sorely in need of a new roof. Thomas rounded up friends in the lower Slocan Valley and had the roof up in a weekend, after which Barlee offered him a job restoring the city hall. Gene Peterson, Sandon's elder statesman, who had lived and mined around Sandon for most of his life, offered them the use of an empty house he owned, later to become the restaurant known as the Tin Cup. And so, in short order, Thomas and Lilly had a house, and Thomas a job, in Sandon—and written agreements about both. So armed, they left their money with the credit union and returned to California for the winter.

In California, Thomas made furniture and Lilly worked in a Mexican restaurant, but they had no intention of staying. They found the government and police oppressive, several of their friends were killed in Vietnam, and with Lilly now pregnant, they were determined that their child, if a boy, not be raised in the United States. Lilly's family were deeply upset; they had worked hard to give their children the American dream and feared that their daughter had become a communist. Nevertheless, in early May Thomas and Lilly loaded everything they owned but a rocking chair on a truck and headed north to five feet of snow on the ground in Sandon. Gene Peterson gave them the keys to the house, which, it turned out, had six rocking chairs amid filth and clutter. Lilly was soon cleaning, sanding, plastering, and painting; Thomas, when Dad and I met him, was building a woodshed. The job for Bill Barlee had not worked out.

Lilly's baby was due in early September. She and Thomas intended to have the birth in Sandon, but got cold feet when the contractions began and drove to New Denver, where a nurse checked Lilly, assured her that all was normal, and sent them back. A few days later I received a letter from Thomas.[3] Lilly had performed "beautifully." Her birth was an "intense and beautiful experience," the baby, Autumn Maple, was "healthy and alert," and he and Lilly felt "a bit smug for having been able to do it at home." He went on to outline plans for the ranch house roof, which, he said, needed more rafters, new two-by-six ceiling joists (the existing joists were one-by-sixes on five-foot centres) and, because the floor was sagging and the foundation giving way in places, a temporary supporting post to straighten the roof. He had bought the hand-split shakes, a "good buy" for $348, and intended to put them on.

In the early fall of 1972, at the very beginning of the restoration of the ranch house, Thomas had established the work's ongoing rule-of-thumb: nothing was simple, every job entailed something else and was

New ranch house roof completed, family camped on the lawn, summer 1973. The interior was a construction site that bears visited from time to time. We were in the tents. The lighter patch of the roof (upper left), unfinished the previous fall, was completed that spring. Harris family photo

far more complicated than it initially seemed. There was another problem, more specific to the roof. Thomas had discovered that cosmic forces were particularly concentrated in the Slocan Valley, and nowhere more so than on the roof of the old ranch house. They were powerful enough to sharpen dull razor blades, and made the roof ridge an ideal place for meditation. Work on the roof during the fall of 1972 balanced these different needs and opportunities to the point that a shake roof was well laid and almost finished when winter closed in. Thomas covered a small patch at the back just below the ridge with a tarp, and shaked it the following spring. Shortly thereafter, Douglas, my seven-year-old son, and a friend were overheard as follows: "When building a house," said the friend authoritatively, "you start with the foundation." "No," said my son, "you start with the roof."

My discussions with Thomas about the work ahead had proceeded some fair way before either of us said anything about its terms. When I eventually asked him what, how, and when he expected to be paid, Thomas replied that he did not want to be paid. Of course, I replied, he must be paid; he had valuable skills—not to pay him would be unthinkable. Thomas was not convinced, and in retrospect his position was more interesting than mine. A hundred and thirty years before, the English Chartists had railed against wage labour, which they held was undermining the opportunity for small farmers and artisans to sell the product of their own work.[4] Thomas knew nothing of the Chartists but, like them, was searching for another economy. On the other hand, the ranch house was hardly a factory. Eventually we compromised. We opened a joint bank account in which I put funds and Thomas withdrew them. This worked fairly well, with me putting into the account approximately what I thought a given amount of work was worth—an arrangement that had the advantage of replacing a commercial contract with trust, but the considerable potential disadvantage of creating misunderstandings. In general, as the work progressed, both of us increasingly appreciated the advantages of more formal arrangements.

Thomas and Lilly remained in Sandon through the fall and early winter of 1972–73, then, in early February, set off for California to introduce Lilly's parents to Autumn Maple. After US border officials ransacked their vehicle and cut its seats, they made it to California, but were back within two months. Shortly thereafter, my uncle Sandy ploughed space for an ample garden, and later, as the weather warmed up, Thomas made a sleeping platform in the woods just above Slocan Lake. The family was relocating to the Bosun Ranch, and it was increasingly assumed that, were the work on the ranch house sufficiently advanced, Thomas, Lilly, and Autumn Maple would stay there the next winter.

By this time, the interior of the log house had been completely gutted. The wall panelling and pressed tin ceilings had been removed, as well as the partitions between rooms, and the stairs and their narrow walls. The fireplace and chimney column had been stripped, found unstable,

A glimpse of the ranch house yet to be, summer 1973. The floor is clean, the pack rats are gone (for the time being), a little furniture is in place, a few chestnut leaves are in a pot, and there is a first touch of elegance. Harris family photo

and dismantled. All that remained under the new roof were log walls with curious openings, weathered floors with gaping holes (where fireplace and chimney had been), and several battered original windows. But the place was finally clean—tons, it seemed, of pack rat droppings had been carted away—and with a little furniture, a pot, and some imagination it seemed possible to look ahead. Later that summer, I bought a dinner set of Minton porcelain, which I was told had come from an old Kootenay family, in a junk shop in Nelson. My stepmother thought me daft; Muriel said that I foresaw something she had not.

In the summer of 1973, Muriel, the children, and I camped in tents on the lawn. Rachel, our third child, who had been a quiet baby, suddenly learned to crawl. Lilly and Autumn Maple were often around. Friends I had invited kept arriving to the point that Muriel did not want to see another car coming over the hill. From time to time, a black bear rummaged in the ranch house at night. Throughout, the whole place was a work site with unexpected openings and drop-offs, and Rachel objected mightily to her playpen. Much of the work was in the basement, but the interior was also being rebuilt.

Some of the fieldstone walls in the basement had crumbled, and the building's northeast corner had dropped some eighteen inches. Thomas said that the place was held up by memory. The alternative, he held, was to brace the floor beams with massive posts, and to support the posts on large concrete pods dug into the ground. This entailed an enormous amount of work because the building sat on clay hardpan, and chipping it away was like chipping through weak rock. We thought that smaller pods would do—that perhaps Thomas relied on a US army manual on building in permafrost—but Thomas persisted. Holes were dug, concrete mixed in a small electric mixer, pods poured, the building jacked up, and cedar posts put in place. Some basement walls were still crumbling, but Thomas had now secured the building, above and below.

It was immediately clear when the partitions were removed that the whole main floor should remain open. It was less clear where the kitchen should go, what should be done with the various openings made to accommodate a larger building, or how and where to build the stairs. In one scheme, the kitchen was in the bay at the front, the back door where the stove now is and opening to an outside deck. There were sketches and family discussions, pronouncements from friends, but no proper drawings, no professional advice. Eventually we agreed that the gaping hole in the back wall should be filled (logs below, windows above), and that sink and counters should go there. If, as we descended, the stairs turned right rather than left (as they had), then there was room for a stove between stairs and counter. That was also agreed. We wanted the stairs themselves to be as open as possible, and when Thomas came up with a promising sketch, the configuration of the main floor space was approximately set.

The upstairs floor, my father said, had always been unsteady. Thomas proposed two sets of reinforcing beams, both supported at their outer ends by logs in the walls, and in the middle by posts (on either side of the stairs) or by bricks (in the chimney above the fireplace). The log walls themselves needed to be covered. We had removed wallpaper, muslin, and planks, and decided to reuse the planks (over fibreglass insulation) with the rough, kerf-marked side out. The finer, varnished cedar panelling, taken from the vestibule of the larger house, would be reversed as well, and used in the stairwell and on the end walls upstairs. Thomas proposed to hand-plane these kerfed surfaces lightly, and fit them so the kerf lines followed each other across the wall. Moreover, most of the window frames would have to be rebuilt, although the sash could be ordered from a woodworking shop in Nelson. And, of course, there were the stairs.

With the basement behind him, Thomas turned to all of this, and over the next year completed most of it himself. When the beams were going in, however, he needed an assistant, and for much of the summer of 1973 Sam Tichenor also worked on the ranch house. Sam, the American-born son of a Presbyterian minister, had grown up in Lebanon.[5] Leaving the Middle East as a young man in 1967, he would not return to the US because of the draft, and turned instead to Canada. Helped by a Quaker draft-counselling service in Montreal, he went to the US Consulate two weeks after arriving in Canada as a landed immigrant and renounced his American citizenship to avoid breaking US draft law. After two years in Montreal, he and his partner, Leah (she from New York and previously married to a draft dodger), hitchhiked to Vancouver, found temporary work, and fell in with the countercultural crowd and various musical groups (voice, guitar, mandolin, sax). Everyone in their circle, Sam said, wanted to go to the Slocan. He, himself, had long dreamed of getting back to the land. In 1971 he, Leah, and their son, Soleil, moved to Hills, fixed up a small cabin on a friend's land, and settled in—just as Thomas, another son of a Protestant American minister, was creating a place for himself and Lilly in Sandon.

Such, plus at times friends and Muriel's parents, was the cast that converged on the ranch house in the loaded summer of 1973. Nobody slept regularly inside, but we moved a big table from the empty manager's house at the Bosun Mine into the middle of the ranch house floor,

The interior under construction, summer 1973. Thomas, on the ladder, is preparing a fitting for a new ceiling beam; Sam Tichenor and a friend from Toronto are at the table; and Muriel is standing. The table, much refinished, has been a part of the ranch house ever since. Harris family photo

and usually ate there. We had a cook stove on the lawn. Sandy came by almost every day to feed the trout in Loch Colin. Thomas and I rescued some planks from the Alamo mill (it was still possible to drive there) and put a deck on what would become the back porch. Lilly and Thomas's garden flourished on the land Sandy had ploughed—and we would spend the next forty years employing various strategies, including pigs, in an effort to extirpate the comfrey they planted. Eventually there was a sink and running water. The McClary wood stove, rejected from the cabin, fit between the counter and the space reserved for the stairs. Two little girls crawled around, and we managed to keep them from dropping into the basement. A bear made off with Muriel's yogourt maker.

At the end of August, Muriel and I and the children returned to Vancouver, and Thomas, Lilly, and Autumn Maple moved into the ranch house. Thomas brought in a heat stove, sealed off the upstairs, and covered the windows with plastic. Work continued indoors and out. In late November, he wrote to me as follows:

> ACROSS THE MEADOW, around the barn and all along the road the snow is crisscrossed and patterned with foot prints and boot-prints and little toe and tail prints. All the critters making their final search for winter stores. Tonight the owl hooted—the hair on the back of the dog stood straight and the little kitten, a novice at the game of catching mice, turned a respectful ear and eye to the blackness beyond the window.
>
> During the day the lake and sky are two tones of the same gun-metal blue. The clouds hide the mountains behind the palest of shrouds, sometimes blue, sometimes grey, sometimes black and brilliant white.
>
> November. November: an old man in a grey coat with a stop-watch in his hip pocket.
>
> The time during which things can be done outside is drawing to an end. It is hoped that when the time is done the things will also be done. As usual, everything takes twice as long as anticipated.[6]

He had picked up the windows in Nelson, and was soon going to begin installing them. For the moment, with both stoves going, the house was "a bit drafty at best." Lilly and Autumn Maple were leaving soon for California, flying from Castlegar. They would be missed, but it would be easier to fit the windows and do inside work without them. Thomas would probably be available to work on the ranch house the next summer. He would try to accommodate my preference for relatively straight lines (mentioned in one of my letters to him) but clearly considered it ridiculous.

Lilly and Autumn Maple returned, but in midwinter the ranch house was hardly habitable and Lilly's relationship with Thomas was strained; she and her daughter went to California, then Hawaii, and stayed there for a couple of months. Later, she would tell me that she knew she would leave Thomas, but came back to complete the house. She had a sister in Hawaii and thought she could remake her life there—without a man if necessary. She also had papers for New Zealand.

When Muriel, the kids, and I returned to the ranch house in the early summer of 1974, the stairs and windows were in, the walls were panelled, and the framed projection along part of the front wall was complete. Although the fireplace and brick chimney remained to be built, for the first time the house looked approximately as it now does. The garden was abundant. Thomas (primarily in the house) and Lilly (primarily in the garden) had done an enormous amount of work. But clearly their relationship was breaking down. They stayed through that summer, then left, and we would not see them again for years. Thomas gave me a carving of an apple tree.

We had liked and admired them both, and they had become good friends although, at a certain level, we were very different. I spent many hours tucked away with my notes and writings in the old ice house by Loch Colin. Thomas was skeptical, and told me one day that if I really knew the stone he held outstretched in his hand I would know all I needed to know. Years before, Ralph Waldo Emerson, the New England transcendentalist, had said much the same, as have mystics through the ages. Thomas reported (long before cellphones) that he at the ranch house and Lilly several hundred feet away at the beach talked with each other. He saw body auras and would tell me, when we differed, that my third eye was showing—an unfair debating strategy, I thought. Increasingly he seemed to see himself as a guru, and earthly attachments as impediments on his spiritual journey. Immersed as we were in family and place, Muriel and I were on a very different track. But there was no gainsaying Thomas's huge contribution to the ranch house. He had taken on a ruin and made it livable and attractive. He had done what he said he would. When he and Lilly left, they were greatly missed. For her part, Lilly would say years later that the ranch had been a spiritual haven, and that it, Sandy, and Gene Peterson in Sandon had set the course of her subsequent life.

The house Thomas and Lilly left behind was quite habitable in summer but hardly finished. The fireplace and chimney were not yet built, some windows were still covered with plastic, and there were many small gaps around doors and windows. Before Muriel, the kids, and I left in August 1974, Gretchen Jordan, who had recently sold her bulk organic food store, the Apple Tree, in New Denver, asked me if she and her two small boys, Aaron and Ischi, might stay in the house that winter. Gretchen was an artist, a weaver, and the terms, to which she agreed, were that she weave us something

Gretchen came from a conservative, right-wing family (Evangelical United Brethren) in rural Pennsylvania,[7] studied at Philadelphia Museum School of Art, worked for a time for Eastman Kodak, became increasingly radical, and was disowned by her family. Moving to San Francisco, she enrolled in the San Francisco Art Institute, met a fellow student and conscientious objector, Glenn Jordan, and married him. They soon left for Mexico to get away from what they judged a corrupt political system and "the craziness of everything," only to be disillusioned by the extent of American influence on the indigenous people of Mexico. Back in San Francisco in 1970, even the anti-war movement seemed consumed by hatred. Gretchen was pregnant. They wanted to get away and find a haven for their family and a more spiritual life. Their lawyer suggested Canada; a friend had land in the Slocan. They arrived in a 1955 International pickup truck with a handmade camper and all their possessions. They had one dollar to their name.

Abandoned by their families, they lived in total poverty and Glenn was wanted by the FBI—the circumstances in which they had another child and their marriage collapsed. Gretchen and her boys moved into a small house on the main street of New Denver, and she slowly re-established the organic food store she and Glenn had begun in Rosebery, the first in the Kootenays. The Apple Tree became a meeting place for old-timers, Japanese Canadians (the store stocked Japanese foods), and hippies. Bush people could pick up their mail and have a bath. When, after three years, the work became overwhelming for a single mother of two, Gretchen sold the Apple Tree and came to live in the ranch house.

Gretchen's ranch house, ca. 1974–75. Gretchen's talent for comfort and coziness is apparent. Carpets, potted plants, weavings, resurrected furniture, and a cat had domesticated the place. Harris family photo

Gretchen was the first to make the restored house a home. She had a feel for coziness, and an eye for attractive arrangements. With a few plants, several worn rugs, some rescued furniture, and a cat, the old house began to look comfortable, but was anything but snug. Gretchen cut and split most of the five cords of wood she burned that winter and spent most of her time tending the stoves. Even so, the house was cold; she bought an electric blanket and huddled with Aaron and Ischi through many a bitter night. After a snowfall, there were always fresh cougar tracks around the house. She kept a close watch on her boys, and saw a lot of Sandy. He kept the road open, and a protective eye on her. Like Lilly, she adored him. For all the cold, she maintains that she loved being on the ranch. Standing by the hay barn on a cold winter night, she could look across fields to shimmering northern lights, which she depicted in the weaving she left as payment. It still hangs upstairs.

Thomas and Lilly, and then Gretchen and her boys, had managed to survive, just, in the old house in winter. I had a sabbatical coming and Muriel and I, together with our three, thought to spend part of 1976 there. Was I to get any academic work done, the house had to be made more habitable.

For one thing, the fireplace and chimney needed to be completely rebuilt. This, it turned out, would be the work of David Lehton, who arrived on the ranch looking for work in the summer of 1975. David was a Californian from near San Diego, a graduate in electrical engineering from San Diego State, an anti-war protestor, and then a draft dodger. From the anti-war tumult of San Francisco, he had retreated to a shack in the California mountains, and had managed, briefly, to obtain conscientious objector status. When he lost it and was called up, he sold his shack and car, hitchhiked north, met a Canadian woman, and crossed into Canada. In 1971 they married in Vernon. Finding "long-haired people" and the Okanagan an uncongenial mix, they moved to Nelson, where, David would later say, the Doukhobors were particularly friendly. Eventually they settled in Rosebery and New Denver.[8]

David had heard that I needed a fireplace, and asked if he could build it. I asked whether he had built fireplaces and laid bricks. He had done neither, he said, but was familiar with cement. I was very dubious. He asked what type of fireplace we wanted. I explained that we intended to use bricks salvaged from the original fireplace and chimney, but rather than the green tiles, oak pillars, elaborate mantel, and mirror that had covered the original (see photo on page 65 in chapter 3), we wanted an unadorned fireplace with the simplest possible mantel. Moreover, we wanted a fireplace after the style of Count Rumford, a late-eighteenth- and early-nineteenth-century theorist of fireplaces. Rumford fireplaces are high and shallow with sharply angled coves (sides) to reflect heat. I gave David the specifications, and he thought he could build one. I remained dubious.

To my surprise, he was back the next day with a detailed, three-dimensional drawing that, brick by brick, anticipated the fireplace now in the ranch house. That settled it. If he could come up with such a drawing,

David Lehton's fireplace and Ken Fagan's stove. The stove was designed to fit beside the fireplace and, except in the coldest weather, it easily heats the house. Note Bob McKay's cupboards on either side of the door and, on a shelf on the far wall, some of the Minton porcelain I bought in a Nelson junk shop. Harris family photo

he could build a fireplace, and he did. It has worked without smoking ever since, although on cold nights the heat it throws out is more than balanced by cold air drawn in from outside. That, however, is in the nature of open fireplaces. David did a fine piece of work, and went on to re-chink all the exterior log work, another major undertaking.

When Gretchen and her boys left in the early summer of 1975, the upstairs was also unfinished. That summer, I panelled the insides of the dormer windows, and when a Vancouver neighbour arrived with siding from his demolished garage, we panelled the angled edges of the ceiling. Bob McKay, another refugee from the United States, and I covered the horizontal ceiling with siding retrieved from the ranch house.

Bob, who would play a considerable role in the house, was an engaging personality and a versatile woodworker. Among many talents, he played the flute, and his background, when eventually I knew it, was impressive and fraught.[9] He grew up in an evangelical Christian family in upstate New York, attended a local college, transferred to Boston University, graduated in chemistry magna cum laude in 1967, married, and planned to go to MIT or Stanford. Changes in the draft law made him eligible for the draft following another degree, but a Rockefeller Fellowship enabled him to enroll, all expenses paid, in the Harvard Divinity School, engage in social activism, and defer the draft indefinitely. He worked with runaway kids at a halfway house and in the courts, and represented the theological seminaries in anti-war demonstrations. When, assuming their status inviolate, he and others in his class sent in their draft cards, they were immediately drafted. The previous year Bob had received landed immigrant status in Canada, and in December 1968 he and his wife, Sue, fled to Canada to escape the draft.

They went to Calgary, where Sue had a contact, then to Banff and various jobs, then, in the fall of 1969, to a group of draft dodgers at Johnsons Landing on Kootenay Lake. Driving one day to New Denver in their old VW van, they met Bay Herman, also a refugee from the United States, in the laundromat. She recommended the community of Hills at the north end of Slocan Lake. Bob and Sue liked what they saw of Hills, bought (with borrowed money) a little land there, then spent the winter in a cabin, rented for ten dollars a month, in Argenta at the head of Kootenay Lake. Their daughter, Aura, was born in Nelson that winter. Back in Hills the next spring they raised a hundred chickens, ate a dozen eggs a day, and Bob earned forty-two dollars a month shovelling lump coal from coal cars for a dollar a ton. They slipped back to Boston in the fall of 1971, separated, and Sue and Aura returned quickly to the Slocan. Bob followed in 1973, this time with some money after building a house in Boston. When he and I worked on the ranch house ceiling, he had been back in the Slocan for two years, working mainly as a carpenter. He designed the octagonal elfin house next to the old Bank of Montreal.

As well as the ceiling, Bob finished the upstairs panelling (following Thomas's use of light, hand-planed boards with undulating kerf lines), gyprocked part of the bathroom, and tightened some of the windows. He built the kitchen cupboards, cutting them from old doors, and made handles from thin rounds of yew wood, cut lengthwise. He made a bench to go with the table from the manager's house (by now sanded and varnished on top and painted white below). His bench was a creation, but too organic for our taste, and he made the two still in use, each with a top salvaged from the Geography Department at UBC. For a time, Bob, Ruth Pal, then his partner, and Aura, his daughter, lived in an old mine building on the Bosun dump with the unobstructed panorama of Slocan Lake before them.

With these changes, the old house became relatively habitable. We moved in at the end of December and stayed into the following summer. The boys went to school in New Denver, I worked upstairs with my back to the chimney, and Rachel, who had seen a theatrical adaptation of James Thurber's *The 13 Clocks* before we left Vancouver, fancied herself the Princess Saralinda. She sat for hours on end in the middle of the floor and repeated over and over to her mother (the cold duke) in a pale voice, "I wish him well." Again, there were cougar tracks in fresh snow around the house.

Sometime that spring, we got in touch with Ken Fagan, an ironworker whose workshop was just off the highway at Lemon Creek, and commissioned a heat stove. We never knew very much about Ken, but understood that he had been a pipe welder in Montana. Certainly he was in California for a time, as he and Thomas Wright had met there. When, and in what circumstances, he came to the Slocan, I never knew. Whatever his background, he was a remarkable artist. He told me that he had lost a commission for a large iron sculpture in the new Vancouver courthouse when the NDP came to power and the building was redesigned. His work was as fluid and organic as plate iron would allow. For a time he made huge lotus blossoms, and when they did not sell he made stoves that looked like lotus blossoms. We did not want a lotus blossom stove, and the design he came up with was a compromise that had no right angles, and almost no straight lines, yet looked tailored. It was intended to fit beside David Lehton's fireplace and connect to a jack in its facade. When Ken delivered his stove, I thought it a masterpiece, and still do. And it works.

Our half-winter and spring in the ranch house showed that it could be lived in fairly comfortably year-round. Moreover, it was basically secure, but an old house has more than its share of vulnerabilities, and over the next several years it was Glenn Jordan, Gretchen's ex-husband, who addressed many of them.

Glenn Jordan, from a Scots-Irish background in rural, southern Alabama, grew up on a small farm that raised corn or cotton and hogs, and brought in seasonal African-American workers.[10] His family was poor—a well, an outhouse, an icebox. Unable after the war to make a go of farming, his father rented his land to a neighbour, did odd jobs, and ran a small grocery, in which, from an early age, Glenn worked. He put himself through junior college and a year in a teachers' college, worked for a year in his father's store, then was accepted for a major in painting at the San Francisco Art Institute. In San Francisco, he supported himself by working as a janitor, encountered the full energy of the early counterculture, and became increasingly radicalized. Drafted in late 1968, he refused on conscientious grounds; his review board in Alabama, in rejecting his refusal, held that young Americans fought for God and country. Glenn was fingerprinted and sent back to California where, after a trial and long-delayed judge's decision spread over two and a half years, he was found guilty. He faced six months in prison and two years of community service. Meanwhile, he had met Gretchen at art school, and the two of them fled to Mexico. Back a year later and advised to go to Canada, they arrived at the border with nothing but their old truck and a few possessions. Canadian border officials advised them to stay in Washington state for a time, get their story together, and return. They did, were admitted on visitors' visas in the spring of 1971, and arrived in the Slocan in the vulnerable and penniless condition I have already described.

They stayed in the Rosebery campsite through the summer, then built a camp above Rosebery, all the while living on the edge and, eventually, out-staying their visas. When an RCMP officer came looking for Glenn, they thought they had been found out. It turned out that his mother had died. Glenn went back for the funeral, was in an automobile accident, and expected to be busted by the FBI. However, as the accident was not his fault, the insurance company offered him a thousand dollars if he would sign a disclaimer, which he did. When he returned, he was able to tell Canadian border officials that he and Gretchen had money for a health food store. They became landed immigrants. When their marriage broke up during the winter of 1972, Gretchen moved the health food store to New Denver, and Glenn went to Vancouver where, for a couple of years, he taught art at a centre for the arts in Burnaby. He was back in New Denver in 1975, supporting himself as a carpenter.

Glenn became the wise, essential backup for an old, ruined house brought back to life. He would tell us what, in our shoes, he would do, and we relied on his judgment. Spread over a good many years, his contributions were as large as Thomas's: baseboards, rebuilt bay windows, concrete steps toward the outdoor bath, concrete basement floor, plywood main floor, back steps and railing, woodshed along the southeast wall, countertops, box benches for storage. He suggested, but did not make, the double-glazed windows we put in the ranch house, built stairs for the cathedral (the old carriage shed), fixed its sagging roof, and built the washhouse. In retrospect, the engagement of the American counterculture with the old ranch house on the Bosun Ranch began with Thomas Wright and ended with Glenn Jordan.

It had been a remarkable time. The ranch house had absorbed just an edge of the vast, far-reaching reaction of young Americans to their country's war in Vietnam. Those who worked on the ranch house had fled the United States, arrived in the Slocan with little or no money, and faced the prospect of making new lives in difficult and radically unfamiliar circumstances. For them, as for so many others throughout the history of modern Canadian settlement, the line between getting by and not was

The ranch house, 1985. By this date the house was largely rebuilt. Muriel's parents are at the right, my father is under the hat, I am standing to his left, my son Colin is up the ladder, and my daughter, Rachel, and two friends are on the porch. Harris family photo

narrow. Basically, their youth and intelligence, coupled with a measure of mutual aid within a loose countercultural community, pulled them through. They were quick and creative learners, and the ranch house was a creative opportunity. Although I and my family were always involved, they did the work, and in so doing put their stamp on a rescued, reworked, and much enjoyed version of the log house my grandfather built in 1897.

Besides restoring a house, the hippies brought their values, some familiar to us, others not. With a background in Fabian socialism and middle-class Edwardian social mores, the Harris family was not accustomed to their suspicion of government, somewhat mystical and frequently drug-assisted spirituality, changing sexual partners, and emphasis on individual freedom. My grandfather had often said that an undue emphasis on freedom was the overriding American problem. Early in my encounters with hippie culture, I often mused that it was another form of classic American liberalism. I am less convinced now. The disrupted circumstances of hippie life made the creation and maintenance of community, on which some hippies expended a great deal of effort, exceedingly difficult. Certainly they shared the belief, always in the ranch air, that the Slocan Valley might offer an alternative to the corrupt and decadent ways of a society left behind. They shared the ranch critique of corporate capitalism and the culture of greed, likewise its appreciation of simple lives lived close to nature. They shared its appreciation of art. Overall, I think, the meeting of the American counterculture and the old ranch house tended to reinvigorate the social critique embedded in the Bosun Ranch. It redefined the ranch—as on another scale it did much of the West Kootenay—as a place where lives might be lived creatively, close to the land, and apart from the mainstream.

12 AFTER THE SECOND GENERATION

BY THE MID-1970S, WHEN THE RANCH HOUSE WAS AGAIN HABITABLE, BOTH MY sister's family and mine spent parts of our summers there. Dad and Marjory, my stepmother, were in the cabin above Loch Colin. John Anderson, a mining engineer, left Kennecott Copper, which had wanted to move him to Costa Rica, and he and my cousin Nancy moved in with Sandy. Mollie, whose memory was gone, was in a nursing home in Vernon. John began running cattle on the Far Field, fattening them for sale, and he and Nancy irrigated some of the hay fields. The cast had changed somewhat, but there were still those who lived on the ranch, and those who came for the summer—and the tensions these differences provoked. Moreover, Marjory, whose background was in business and whose father, H. H. Stevens, had been a prominent Conservative politician, brought very different values and aesthetics, which, located in the Loch Colin cabin of all places, I found very hard to deal with. And, although John Anderson much admired his father-in-law, the two of them, living in the same house, did not always get along. There were times during these years when I found the ranch unsettling and dark.

Also by the mid-1970s, it was clear that my father's memory was failing. Sandy's memory was better, but he too retreated into his favourite stories, which Nancy and John heard hundreds of times. Marjory, to her great credit, looked after my father as long as she could, but eventually he had to be placed in a nursing home, where he died in 1989. His will gave Marjory the right to use the cabin as long as she wished, but left it to my sister. He assumed the ranch house was for me, and left his two lots, which comprised some thirty acres near the southern boundary of the ranch, to us both. One

Cleaning out the dam, summer 2008. Thomson (foreground), Éric Taillefer (looking over dam), Rachel (with shovel), Candy (second from right), then Douglas. Note the gendered division of labour. Harris family photo

of these lots was a product of the division of the ranch between Sandy and my father in 1964 (see chapter 9), and the other of a land swap circa 1972. The latter, the Bosun Mine property (11.2 acres) had been on the market for years, and eventually my mother, suspecting that the mine property would find a family use, had bought it. A decade later, Sandy and my father agreed that, as it divided Sandy's fields in two, it would be given to him in return for equivalent acreage adjacent to my father's lot. My mother had been right. This trade created a viable space around the old ranch house. The cabin was another matter. Marjory returned in summer for two or three years after my father's death, and then passed it on to my sister. My sister and her family, who had also used and loved the ranch house, were now established in the cabin; Loch Colin and the field below the cabin became their sphere of influence. I and my family were in the ranch house; the adjacent farm buildings and fields were our sphere.

When Sandy died in 1986, he left most of the ranch to his daughter, Nancy. John was now the principal male on a ranch of which he owned

Colin, Rachel, and a friend in Thomas Wright's outdoor bath, July 1978. During the war the elderly Japanese Canadian men in the ranch house pickled cabbages in the large wooden box behind the bath. Harris family photo

nothing, and Nancy quickly converted her ownership of the ranch into a joint tenancy. On the death of one, his or her share would pass to the other. John was active in the community in these years—on the regional district board, on the school board, an unsuccessful candidate for mayor of Silverton—but his base was the ranch, which increasingly he thought of as the Anderson Ranch. There he inherited, and intensified, the negative feelings about the sojourning Harrises that had run through the ranch for years. He began to talk of the ranch as a prison in which the A. L. Harrises and now he and Nancy worked while others came and played. He could be charming or explosive, and most of us became wary of him. Yet the question that no one talked about was still in the air: what would happen to the ranch when John and Nancy were no longer around? On one occasion, I wrote to them, offering to buy the ranch on the understanding that they would have unfettered use of it throughout their lifetimes, after which it would revert to me or my heirs. When I talked with them, John was dismissive and angry. I had touched too many raw

In the old apple tree, summer 2005. This tree, a Yellow Transparent, is the lone survivor from the original orchard planted almost 120 years ago. Harris family photo

nerves, particularly that they were childless and that, in his eyes, the freeloading Harrises did not deserve the ranch.

By the mid-1990s Nancy's memory was also failing. Nancy had always admired her father, and in many ways was much like him. She had grown up on the ranch, then studied zoology and botany at UBC. Although much more scientifically trained than her father, her focus was the same nature that had increasingly captivated him. For years she organized the annual camps of the BC Natural History Society. She knew many remote corners of the province, and worked with some of its best natural scientists. She and her friend Dr. Bert Brink laid out many of the province's ecological reserves. From my perspective, she was a terrific companion on any alpine hike because she knew the flowers, the birds, the alpine ecology. I enjoyed and admired her; as kids we climbed the New Denver glacier, battling through alder slide with heavy packs long before anyone

In the willow tree, summer 2013. This tree, planted seventeen years ago and now far larger than the 120-year-old Yellow Transparent, thrives at the edge of a marsh. Harris family photo

rebuilt the trail. But as her memory declined, all of this began to slip away. She stayed much closer to home, and relied increasingly on John. For his part, he took on more and more of the housework, and more of their talk became ranch banter. John, whose memory was excellent, was increasingly alone.

He was in a difficult position, which may partly explain why he railed so against the fair-weather Harrises. For my part, his and Sandy's critique of us always seemed partly justified. We did not and could not live on the ranch. Although Muriel and I and our three children once spent eight months there (while I was writing) and, years later, our son Douglas, his wife, Candy, and their two-year-old son moved in and lived there for a year and a half (while Douglas was finishing a thesis), we did not know the ranch as Sandy did. With the exception of Steve Pond, my brother-in-law, we were not particularly practical. None of this was easily changed. We helped as we could when we were there, but probably not very usefully. We always had projects, but they usually had to do with our buildings and

On the raft in Loch Colin, summer 2003. This pond, made with horse and dragline in about 1907, was intended for ducks. Harris family photo

close surroundings. We reamed out the dam on Harris Creek, an annual job, and paid for a new water pipe from the dam across most of the ranch. We were not indifferent to the larger ranch and its needs, but were not there through most of most years.

From 1971, Muriel and I and our three children, Douglas, Colin, and Rachel, lived in Vancouver. When our children were grown up, they all moved away, Rachel at seventeen to study modern dance in Montreal. Only Douglas and his wife, Candy Thomson, returned to stay. Their children—Thomson, Molly, and Ellen—would grow up in Vancouver. Colin and his wife, Janet Chow, settled in Ottawa (Janet grew up in the Ottawa Valley), and raised their children—Rowan, Teagan, and Griffin—there. Rachel stayed in Montreal, married a Quebecois, Éric Taillefer, and continued dancing while raising two daughters, Lily and Alice. In various combinations, this crew descended upon the ranch as often as time and finances allowed, but none of us lived there.

Perhaps we were sojourners, but we loved the ranch, and on our terms used it well. While Douglas or I often retreated to write, overall it became a place for family and friends to gather and enjoy its beauty and each other. It was full of talk and odd doings, including an outdoor bath, which Thomas Wright had set up, with a fire underneath and a hose for

On the raft in Loch Colin, summer 2005. Effort is not necessarily rewarded. Harris family photo

a thermostat. For kids, there were trees to climb, Loch Colin to play in, and endless opportunities for messing about. The beach and Slocan Lake were a few minutes away, down a road built before 1900 to connect the ranch to the lake steamer. There were fishing trips to the local creeks, huckleberry-picking expeditions, and mountain hikes. Although we were not much good at it, we often read plays, sometimes ones that father had directed at John Oliver High School, sometimes Shaw or Shakespeare. More often, a parent or grandparent read to children. There was a lot of music, and in Rachel's time, dance performances on the lawn or fields. And there was always wood to chop, a garden to weed, walks, and conversations. Much of this drew on what my sister and I had known at the cabin, but my wife, Muriel, also encountered that world the summer we met, and in her way became as welcoming as my mother. In these ways we used the ranch, and it deeply penetrated our lives. For most of us, we had come from there.

Messing around, summer 2010. The building is a chicken coop; the sculpture (left) is a bug made from old farm equipment. Harris family photo

A reading, summer 2007. Colin with a young audience. Harris family photo

Music at Douglas and Candy's wedding, August 1997. From left: my sister, Susan, her son James, two friends from Vancouver, and Rachel. Harris family photo

Rachel in performance, summer 2010. Such performances for a local audience have become annual events. Tom Perry photo

13 THE CLAY HOUSE

WHEREAS THE RANCH HOUSE WAS RESTORED IN THE 1970S, THE CLAY HOUSE was built from scratch more than thirty years later by another group who found a haven in the Slocan. Times were very different. The Vietnam War was long over, and the war in Iraq had not created the same electricity. The fervour of the sixties and early seventies was gone. Many of the hippies had left the valley, and those who remained lived much-tempered lives. Other people had come, many of them Canadians accustomed to rural life at the edge of the bush and seeking a quiet Slocan niche where they could live sustainable local lives apart from a materialistic and urbanized world. Some of them became our good friends, and the ranch, with their coming, became more used and lively. They became as central to the clay house as the hippies had been to the ranch house.

Their involvement began this way. My daughter, Rachel, got to know Julia Greenlaw at a Friday market in New Denver in mid-July 2007, and invited her to tea. Julia, her husband, and two young sons came by two days later, the day after Flor, Rachel's stepdaughter, who happened to be looking out the living room window, saw the ancient chestnut tree at the gate fall. There was no wind; the massive trunk just dropped. When Julia and her family arrived, a tangle of chestnut obliterated the lawn. Julia's husband produced a chainsaw and, wielding it within inches of his open-toed sandals, quickly reduced the mass to order, then loaded the leafier branches atop an old Toyota Tercel and drove off to feed his goats.

That is how I met Norbert Duerichen, a first encounter that was not anomalous. His goats preoccupied him, and he was adept at finding uses for whatever was at hand. He was born in Austria late in the war, and his family moved to Quebec in 1949, then a year later to a farm seventeen miles from Smithers at a northern edge of North American agriculture. His parents were holistic biodynamic gardeners (after Rudolf Steiner) at a time when hardly anyone in northern British Columbia had heard of such ways. With neither running water nor electricity, the Duerichen farm was a struggle. By the time they were in their early teens, Norbert and his older brother, Hans, did much of the farm work and were known for their mechanical skills. Eventually, Hans became a mechanical engineer, and Norbert a builder, something of a jack of all trades, a creator of ingenious devices, and a critic of modern ways. When I met him, he and his family had recently left an increasingly crowded Galiano Island (off the southeast coast of Vancouver Island), and moved to the Slocan.

Norbert distrusted mechanized commercial agriculture and its equipment. He felt that the high point of agricultural technology had been reached in the 1950s, when machinery could still be repaired on the farm. He admired the accumulated wisdom embedded in traditional rural economies, and considered it wholesome to live as locally as possible. Hence the goats. They provide milk, manure, and, as they browse adjacent woodland, reduce the fire hazard. Had he more space, he'd have kept a horse. More than anyone I had known, he had the skills needed to live comfortably on a small farm at the edge of a forest in western Canada. In relation to his, my specialized academic life seemed bounded and narrow.

One of his many enthusiasms was for a type of wall construction that surrounded a wood frame with clay and wood chips. He had a German manual on the subject—*Leichtlehmbau: Alter Baustoff – Neue Technik*—and extolled the method's advantages: cheap local materials, high insulating values, fire resistance, durability, and the satisfaction of making it. I had quickly become very fond of Norbert, and greatly admired his practical ingenuity. And so one day, after listening to an exposition of the virtues of light clay construction, and without thinking much, I said, "Well Norbert, what do you say to making such a house, a small one, on the ranch?" We soon agreed that construction would begin next spring. My wife, Muriel, and I had discussed the need for more accommodation from time to time, but not more. The house, she said, just seemed to happen.

Certainly, there was a need. When Muriel and I, our three children, their families, and assorted friends assembled at the ranch, we no longer fit. Moreover, we wanted a house that could be lived in easily year-round, as the ranch house barely could. But it was Norbert who got us going. Without him, the house would not have been built, not when or as it was.

Muriel and I settled on an elevated, well-drained site on a low terrace north of the main barn. It overlooked the ranch's most productive field, the clutch of old buildings that comprised the heart of the ranch, and a panorama of mountains. It was far enough away from the ranch house to be independent, close enough to be involved. It seemed symbolically right: farmland before, forest behind, a Kootenay example of the pinched transcontinental span of Canadian settlement. The house itself was to be small, efficient, and easy to live in year-round. It was to use local materials as much as possible, be fun to build, and have a certain elegance.

I began to design it in early October 2007. In a first crude sketch, there is no second storey, the bathroom is in the middle of the back, a bedroom to one side, a mud room to the other. There was a wide porch along the facade and the north gable end, and a smaller porch, intended for a summer kitchen, at the south end. A week later, the ground floor plan assumed approximately its final configuration. The summer kitchen remained and there was still no second storey, but we realized that floor space could be almost doubled by raising the height of the side walls a couple of feet. A December plan shows a second storey accessible by stepladder from the mud room.

It remained to work out the access to the second storey and the arrangement of windows and doors. I favoured small windows. The house would be easier to heat in winter and keep cool in summer; moreover there were views aplenty outside. The rest of the family wanted bigger windows. By May 2008, the back porch had been enlarged and the south end porch for a proposed summer kitchen had been dropped (for reasons we can't remember), but the windows remained under discussion, and final decisions emerged only as construction proceeded. On Colin's advice, the upstairs windows on the north gable, and on Éric's, the north window in the main downstairs room, were enlarged. Mary Subedar and Knut Haugsoen, architect friends from Winnipeg, arrived during the downstairs wall framing, recommended a vertical window in the stairwell, and made a sketch of the stairs, suggestions we largely followed. Overall, I made initial proposals and defended them with diminishing effect.

Design was far from complete when construction began in the spring of 2008. First, power had to reach the work site, and we wanted the line underground. In early April, Norbert got wind of second-hand electric cable from an oilrig in McBride on the Yellowhead Highway 125 miles east of Prince George, and he and Julia set off to fetch it. It was far heavier than any two could lift, and Norbert designed a spool, wound the cable around it, and then contrived to pay out the cable into the ditch together with orange telephone pipe. With a backhoe at hand, we also put in a septic tank and field. Then Jim Fitchett, whose grandparents had come from England to the bald prairie near Drumheller, Alberta, in 1910 and somehow survived in a sod house, arrived with an excavator and managed to eat away enough shale from a mound that was almost bedrock to create a crawl space. Norbert had reservations about concrete, but relaxed his principles. By late June the foundation was framed, cement poured, and the first floor joists laid.

By this time, Norbert had assembled a building crew, the principals of which were Chris Bokstrom and Jeff Pilsner. They were in-laws of a sort—Jeff lived with Chris's wife, Ana, from whom Chris had been separated for a decade. Both had withdrawn from the world of getting and spending and, together with Ana and her two sons (by Chris), had settled recently in the Slocan. Chris, son of a Swedish father and English mother, grew up in Pemberton and North Vancouver, studied forestry and engineering, and for a time owned and managed a forestry consulting company in northern Alberta. Jeff, whose father was Austrian and mother was from French Louisiana, grew up in Regina in a family dominated by the apocalyptic vision of the Worldwide Church of God. After high school, he

With a crawl space excavated and foundations poured, Chris Bokstrom (left) and Jeff Pilsner began work on the floor joists, June 2008. Harris family photo

entered the gilded world of Ambassador College, an evangelical Bible college in Dallas, Texas. Eventually, and traumatically, he left the Worldwide Church of God, and the church, riven by scandal, collapsed, leaving Jeff a bruised survivor. He worked in sales, marketing, and publishing, while singing and acting with choral and theatrical groups, then met Ana in Edmonton at a self-improvement course. In the summer of 2006, after many wanderings, the five of them fetched up in a dilapidated cabin not far from the Bosun Ranch. Jeff, Ana, and the boys lived in the cabin, Chris in his trailer.

This was the crew—Norbert, Chris, and Jeff—that laid the clay house floor and framed the walls and roof. Norbert was the leader and by far the most experienced builder, but Chris was also a skilled problem solver—and all three were full of political opinion. Essentially, they were gentle anarchists with an apocalyptic vision of global meltdown in an urban-industrial civilization running out of control. They had no use for big business and not much more for government, and held that the world at

Framing the clay house, July 2008. The basic construction materials, clay and wood chips, were not load-bearing, and were applied around a wooden frame. A crane placed the central roof beam (cut on the ranch); Norbert fitted the knee joints. Harris family photo

large was recklessly off track. They were vocal critics of global consumerism, and of the current pace of development and consumption. Norbert's vision was the most retrospective, Jeff's the most tuned to contemporary technology, but all three sought to live locally and gently with nature and society. They were not hippies. They brought practical experience, and Norbert and Chris especially had lived for much of their lives at the edge of the bush. They had all been drawn to the Slocan as a place somewhat apart, where people and nature might live well with each other.

They built quickly. The house was hung on a central roof beam and supporting posts, all cut and milled on the ranch. The framed walls, built of dimensioned lumber, consisted of inner two-by-fours and outer two-by-twos. Cross-braced one-by-six planks nailed to the inside and outside of these studs created a space, twelve inches wide, for the clay and wood

Norbert and helpers mixing clay and water in a drum, August 2008. The resulting clay slurry was mixed with wood chips in the angled drum mounted on tires. Turning slowly, it deposited the pile of clay-coated wood chips that Douglas is raking. Harris family photo

Breaking up clay, August 2008. Thomson, Teagan, and Rowan are working lumpy clay through a sieve, Norbert is tending the mixer, and I am looking on. Harris family photo

shavings, both obtained in Nakusp, that would fill in the wall. Before combining the two, lumps in the clay had to be broken down. Then fine clay and water were mixed in a large barrel with an electric propeller clamped to its side. After this, a clay slurry and wood shavings were added to the upper end of an open-ended barrel placed diagonally on tires on the axles of a Ford Ranger pickup and a trailer. Turned by a one-horsepower electric motor and geared down, the barrel rotated at roughly the speed of a cement mixer. Inside, baffles prevented the mix from running out too quickly. The whole contraption, which looked as if someone had planned a moon shot in a junkyard, was Norbert's doing, and it worked. The pile of clay-coated shavings dropped from the barrel cured for a few hours, then were tamped into the space between the planks nailed to the studs. Two plank widths (twelve vertical inches) around the whole house was a day's work. As the walls rose, Chris and Jeff did most of this clay work, but my son-in-law, Éric, worked for hours to break up clay, as did Rowan, Thomson, and Teagan. Even Molly and Griffin worked clay.

Rachel and I had selected the metal roofing, which by this time was in place. A wood stove could now be put in. Norbert recommended a Mennonite stove sold by Berry Hill Farms in Ontario, and I ordered one. When I phoned to enquire about the delivery date, I was told that, as the Mennonites had no phones, the best Berry Hill could do was contact a local farmer who might then contact the Mennonites. The system seemed frail, but the stove arrived three weeks later. With a chimney, and plastic on the windows, we now had a house that could be heated. The floor and roof were insulated (with blue cotton looking like diced blue jeans), and the plumbing was in, but there was the matter of covering the underfloor insulation with wire mesh to protect it from mice. At this point Ana Bokstrom, Jeff's partner, whom then I hardly knew, arrived and disappeared into a tight crawl space above shattered shale, from which, in effect, she emerged after five days on her back. The insulation was protected, even around the plumbing. Ana seemed remarkable.

A building season was almost over. We had not got a building permit. Nothing on the ranch had ever been built with one and we had maintained

Muriel putting clay-coated wood chips in a frame, August 2008. Note the pairs of braced studs that comprise the support for the walls, and the frames around the lower walls into which Muriel is packing the product of Norbert's mixings. Harris family photo

Jeff (left) and Chris (right) working clay, August 2008. Two rounds of framing boards could be raised each working day. As the frames got higher, Jeff and Chris took over most of this work. Harris family photo

the tradition. However, within the limits of light clay construction, we had tried to follow the code, and Norbert's brother Hans, a mechanical engineer, had checked drawings of the foundation. But someone spilled the beans, and there was a stop work order on one of the porch posts. Given that the work for the year was largely done, its timing was fortunate, but ahead lay a fine and negotiations.

Peter Southin, the building inspector from Nakusp with whom we dealt in the new year, had some sympathy for alternative housing (he lived in a log building) but administered a building code that ignored clay houses. He thought the porch posts too small, had reservations about the foundation, found the venting around the stove and in the bathroom inadequate, and the clay walls suspect. Some of his objections could be addressed, but we could not change the basic construction, and it became clear that Southin would not pass it. He wanted an engineer's authorization, and it was Norbert's brother, Hans, who cast an engineer's eye over the house and bailed us out. His calculations and engineer's letter, together with payment of fees and fine, secured a building permit.

One of the first jobs of 2009 was to put a concrete deck on the porch, and Gerry Wolfe arrived to lay it. I had employed Gerry before, and knew him to be an excellent cement man and a talker. He considered himself a philosopher, and his philosophical ruminations tended to be lengthy and somewhat baffling. Gerry believed in reincarnation, and was pretty sure that he had once been a Roman centurion, then a mathematician, then a musician, and finally, before his present incarnation, General George Custer. It was an impressive lineage, and, whether it helped him or not, he laid a fine deck.

Ana and I began to build window frames, and as we did, parts of her life slowly emerged. Her mother was of Russian Mennonite stock, her father a Nova Scotian Baptist and a master carpenter. Working frequently with him as she grew up, she learned carpentry, but her formal training was in interior design. Hired by a Vancouver design firm, she designed home interiors for wealthy families. Later, in Ottawa, she was offered the

Molly and Griffin working clay, August 2008. One of the great advantages of light-clay construction is that the work can be widely shared. Harris family photo

position of director of design at the National Gallery but turned it down. Pregnant, she wanted to return to the West. She and Chris went to northern Alberta and started their own forestry company, Chris managing the field operations, Ana the office and the GIS mapping. They had a second son, but their marriage was troubled and they separated in 1998. She met Jeff and enrolled in the Alberta College of Art and Design in Calgary. Jeff looked after the boys until, as they wished, he and Ana took them back to their father. Then, feeling fragile, hating Calgary's rampant consumerism, and wanting to turn their lives around, they fled to a co-op in Ecuador and, hopefully, a simpler life.

Ana and her sister, who had married into one of the wealthiest families in the United States, had completely parted ways. With an annual disposable income in the millions, her sister's husband never worked and lived among the desperately rich. Ana's stories are remarkable—about,

Me on Gerry Wolfe's deck, July 2009. The deck is finished but the windows are not yet in and the walls are not yet plastered. The machine just beyond the deck, another Norbert creation, was intended to press clay-coated wood chips into bricks. This creation did not work. Harris family photo

for example, a friend of her brother-in-law's who rented Dodger Stadium in Los Angeles for a wedding proposal, and, after an elegant dinner for two served at midfield, offered the young woman a choice: either to marry him or take the new Rolls-Royce parked nearby. (She made the wrong choice, Ana said.) Ana and Jeff were on another track. The co-op in Ecuador had not worked out, but their determination to live a simple, alternative life was firm. They stayed in Ecuador for a time, then returned to Canada, bought an old motorhome, and went to California, where Chris, who had wound up the forestry company, and the boys joined them. They became an unusual extended family: the boys with Jeff and Ana in their motorhome, Chris in his tent—an odd but enduring domestic arrangement. Leaving California, the five returned to a winter in a campground on Kootenay Lake, then to two and a half years without electricity or running water near Sayward on Vancouver Island, then, in 2006, to a cabin in the Slocan.

As Ana and I began to work together, I knew none of this; she seemed dauntingly competent. We hadn't worked very long when she offered me a cup of tea, which I managed to sit on. A short while later, she offered me another cup, which I also sat on. Not a good start. The other problem was that the wood we were using, two-by-twelve fir planks milled on the ranch, was cupped and twisted. It was the devil's own job to get a true frame out of such wood, and our wrestling with it for several days produced only two window frames, one in the stairwell, the other in the bathroom (the latter remains a measure of our difficulties). At this point Norbert's step-son, Blake Richards, appeared. Blake, a professional carpenter, credited Norbert for most of his skills, and viewed our efforts with disdain. We were using the wrong wood and a poor set-up. He took over, switched to cedar, and made and installed all the rest of the window frames in a day and a half. But he worked with straight cedar, whereas Ana and I had been determined to use our locally milled fir. There was nothing wrong with Ana's problem-solving skills or with their range which, I was beginning to realize, were different from but equal to Norbert's.

In July 2009, Éric, Chris, and Ana laid the upstairs floor. There were as yet no stairs, only discussions about how best to make them. Eventually, we approximately followed Mary and Knut's design sketch. Using ranch wood, Chris and Norbert made the supporting stringers, treads, and risers, while Ana made the bannister and pales. We toyed with burlap to cover the upstairs insulation, but settled on one-by-six boards. I bought Farrow & Ball paint, but we liked the base coat and left it at that. Ana mixed the Farrow & Ball floor paint I'd also bought with half a litre of dull red to make the paint for the upstairs floor. I put in the tamarack panelling in the stairwell and the main room downstairs, and the cedar panelling in the bedroom.

The larger job was plastering, and the person in the Slocan for this job, I was told, was Cindy Walker, described as an expert in medieval plastering. Cindy was a Gustafson, a descendant of a Swedish couple that

Cindy Walker (left) and Ana Bokstrom (with hat) plastering, summer 2009. Skilled medieval plasterers, which Cindy certainly was, are hard to come by. Harris family photo

settled near Perry Siding (south of Slocan City) more than a hundred years ago. When we got in touch with her, she was living by herself in a tiny, self-built house well off the highway a few miles south of Silverton. Cindy was determined and resolute, a relentless worker. She had honed her plastering skills in New Mexico and Arizona and demanded near perfection of herself and anyone working for her. Her young male assistant held her in apprehensive awe. Ana, who among her other talents was an accomplished potter, was less easily intimidated. She joined the operation, and the plastering proceeded apace, inside and out, for three weeks. The materials—sand, clay, diced straw, and water—were applied with trowels in two coats: a thick undercoat, a much thinner finish coat. It was hard, physical work capped indoors by a final sponging to smooth the plaster. The result: handmade walls with tawny surfaces that undulate slightly in angled light. Cindy knew her trade.

With the building permit, we progressed in these considerable ways, but the stop work order eventually produced two other inspectors. An electrical inspector rejected the cable brought from McBride, and wanted another cable, another trench, and a new transformer. A water inspector accepted our septic tank, but rejected our septic field.

These blows had to be dealt with. Eric Waterfield, the well-named water inspector, asked me to dig test pits, fill them with water, and record the time they took to empty. I dug several in the slope below the house and each emptied in seconds. We ended up with two sixty-five-foot drains from the septic tank running along the slope between the clay house and the field—enough septic capacity, it seemed to me, for a small hotel. But it could not be helped, and Eric Waterfield was satisfied. Faced with the expense of the electrical inspector's proposals, we went solar. The valley's solar expert, Kip Drobish, and his wife, Marcy Mahr, had recently immigrated from Montana, drawn by the premise that, overall, Slocan society was ecologically more sensitive, and he put in the system: two panels on the roof, four batteries in the crawl space, controls in a corner of the mud room. We switched to LED lights, and moved the large fridge, an eyesore, to the barn. Whenever the sun shone (and for a few days after), the system worked.

In early September José Bothello, a local cabinetmaker with a shop near Rosebery, put in the windows and doors. Ana built a temporary kitchen counter, faced it with fabric that I brought from Vancouver, and fashioned a backboard from a knotty plank. She also laid the bathroom tiles. Although the rest of the downstairs floors and the kitchen cabinetry were unfinished, there was now a habitable house, and it was agreed that Jeff and Ana, who had spent the previous winter in a yurt at Rosebery, would live in it. They were there when, on October 31, Muriel and I, jack-o'-lantern in hand, paid them a Halloween visit. With them was their friend Tommy, who, we learned, was a shaman. Apparently Jeff and Ana had been sleeping poorly since they moved in, and Tommy had offered assistance. He had laid out the tools of his trade on the table, and was about to intercede with the local spirits. We and our overmatched

The completed clay house, November 2009. The land around it still bears the roughness of construction, but the house itself is built. The barn, dating from before World War I, is largely a ruin. Harris family photo

jack-o'-lantern withdrew. Later we learned that Tommy contacted very few spirits in the house—it was too new—but found the ranch itself teeming with spirits, almost all of them Japanese. So delighted were they to be contacted that they pestered him for three weeks until he directed them to Mount Fujiyama, from where he assumed they would find their way home. Jeff and Ana remained in the clay house, sleeping soundly enough, through the rest of 2009 and well into 2010.

During the spring and summer of 2010, there was not a lot more to do: tamarack flooring (most from the ranch) in the main room and the downstairs bedroom, tile in the mud room and on the bathroom wall behind the bath (brought from Vancouver and laid by Ana), kitchen cabinetry (made and installed by José Bothello). By the end of the summer the house was complete except for some decisions about lighting. It had cost less than an ordinary house of equivalent size.

It seemed to me that we had done what we set out to do. The house was built largely with local materials, and is simple and easy to live in. It takes pressure off the old ranch house and, beginning in the summer of

The clay house in situ, October 2017. The clay house has mellowed into the landscape, the barn has been restored, and on a Thanksgiving weekend Thomson and Molly, now grown up, are digging potatoes. Harris family photo

2011, our children and their families have loved to be there. The windows that Colin and Éric enlarged, both of which look north across a ranch field to the mountains at the head of Slocan Lake, are particularly enjoyed. I like the feel, smell, and texture of the place, especially when unfurnished. Cindy's untinted and unadorned plaster walls, José's simple tamarack cabinetry, the unfinished cedar beams in the ceiling, and the deeply inset windows have created a gentle, harmonious interior space. Outside, the house feels part of the land, and will even more as the years pass.

I began the design, but the clay house is the product of other talents, particularly of Norbert's in construction and of Ana's in finishing, and those of a group of remarkable people—Chris, Jeff, Cindy, José, and others—and of their shared criticism of contemporary consumer society. Beyond work and skills, they brought the edge of a vision, not unusual in the Slocan or in the tradition of the ranch, of another way to live on this earth. And as we built the clay house, and as different talents and ages contributed as they could, the line between work and fun blurred. Had they seen what was going on, William Morris and my grandfather would probably have smiled.

In the fall of 2010 Chris Bokstrom moved into the house and has been there through subsequent winters. In summer, when the family arrives, he retreats to his tent.

EPILOGUE

AT THE BEGINNING OF THE TWENTY-FIRST CENTURY, THE QUESTION OF THE future of the ranch hung in the air more insistently than ever. Nancy's memory continued to deteriorate, John acquired power of attorney over her affairs, and eventually, when he could no longer look after her, found a place for her in the Pavilion, an intermediate care residence (originally the sanatorium for Japanese Canadians) in New Denver. By this time he was drinking heavily and seriously overweight, and Ralph Wilson, one of John and Nancy's foster children, came to look after him. Ralph loved Nancy, who had provided most of the affection and stability he had known in his life, and once told her that he would look after John. The two men needed each other—Ralph because he had injured his back and could no longer work in the mines, John because he could not look after himself well enough—but, cooped up in Sandy and Mollie's house on the ranch, they barely got along. Nancy had often said that she wanted the ranch to remain as it was, and that it would be left to the Nature Conservancy, but it never was. Now her mind was gone, and it was John, once a promising young mining engineer but now alone, unhappy, and near the end of his life, who controlled the fate of most of the Bosun Ranch.

He had no children, could not abide the thought that the R. C. Harrises might acquire the ranch, and convinced himself that he, like Sandy and Mollie before him, lived in a prison. The conviction that those who lived on the ranch worked while others played became an obsession. Nor was John able to look after the ranch, which still produced a thin crop of hay but essentially was untended and waiting. But waiting for what, for the coyotes, the deer, and other wild creatures? It was in this frame of mind and in the boom real estate years of 2006–7, that John was approached by developers. They assured him that the ranch was ideal for a high-end real estate development—for vacation homes for the rich from around the world, people who had three or four such homes and moved among them. Roads would have to be upgraded and paved, water pumped from the lake, and hydro provided to all lots. But all of this was feasible, and all of it would be exceedingly profitable. The one hundred lots on the lower terrace would sell for $500,000 each. When they were sold, the upper terrace, which had never been cleared and farmed, would be similarly developed. In its first phase, this was a $50-million development.

For John, lonely and unfulfilled, this was an enticing prospect, a capstone on a life that otherwise had not quite worked out, and payment for years of ranch work. He imagined himself in a large house on the brow of the old Bosun Mine property, and planned to give a great deal of money to New Denver—a skating rink, for example, so local boys could aspire to reach the NHL. He would be a respected philanthropist; the ranch would contribute far more to the valley than ever before. With such thoughts in mind, he entered into an agreement with the developers, giving them title to half the land in return for their work to promote and develop it. As he no longer had a majority position, the developers effectively controlled the ranch, except for the two pieces my sister and I owned. Shortly after this agreement Nancy died, her memory completely gone.

For me and my family, these plans seemed a huge rasp drawn across much-loved land. Where was my grandfather's vision of a sharing, cooperative, and far more egalitarian society? Where were his years of work to create a hillside farm producing good, healthy food? Where was the space for Chunky (Sandy's tamed grouse), for Sandy's coyote friends, for his deer? Rather, the ranch seemed about to be turned over to the very parts of society of which the ranch had always been a severe critic—to, in my grandfather's eyes, the useless people. Perhaps land itself has no memory, though I sometimes think it does, but we who loved the ranch remembered, and were appalled.

My sister and I were asked to participate. If we would place our two lots within this bare strata development, the development company would drill each of us a well. However, numbed by its plans, we were in no mood to cooperate. Nor, as it turned out, were many New Denverites. The company's plans to extract and, after treatment, return water to Slocan Lake were seen as ecological threats. Its lot values, if sustained, were considered likely to drive up tax assessments, thereby making it harder for ordinary people to live in the valley. The jet-setting international rich were seen as measures of the world's ills rather than its achievements. Many New Denverites did not want them around. Some who had known Sandy and Nancy thought they must be turning over in their graves. There were many letters in the local newspaper, almost all of them critical of the company's plans. My brother-in-law, Steve Pond, considered the whole scheme a speculative fantasy. I was less confident.

In retrospect, my brother-in-law was right. The scheme ran into the financial collapse of 2008. No lots were sold, and plans slowly withered. The company reduced the one hundred lots to just eighteen, some of them large, but they did not sell either. Eventually the company put the whole property on the market, but at a price far above going rates for Slocan land. There were no takers. At one point, a local group approached the company about a possible sale to the community, but the sum the community thought it could afford was a quarter of the company's price. And so the years passed and the Bosun Ranch sat unsold. In 2013 the company, apparently needing funds to cover mortgage payments and taxes, clear-cut a good portion of the upper ranch. By this time John and his co-owners were barely in touch with each other. John watched TV, became increasingly unhealthy, and probably bitterly regretted a deal that had promised much and returned nothing. Angry and unhappy, he lost touch with almost all of his friends. According to Ralph, he blanked out his deal with the company, read none of its paperwork, and signed whatever it asked him to. In February 2014 he died of congestive heart failure.

As the price of the ranch fell, Muriel and I, and our children, began to consider whether we should try to buy it. The question, essentially, was to what end? At that time, none of our three children could imagine living there in the reasonable future. On the other hand, they and our eight grandchildren were deeply attached to the ranch. Most of them had been there for a good part of the summer almost every year of their lives. The probability was that the ranch would be used, and used well, by the family for years ahead. But it would have to be looked after. There would be decisions and expenses. The ranch could not sit untended. Such were our discussions, although the price still seemed out of reach. Then, suddenly, in the summer of 2015, the company was prepared to sell District Lot 1799—the land it had just logged and the mountainside beyond—at a reasonable price. It was badly damaged land, and not very accessible for others. But for us, it included a crucial field adjacent to our land, the dam on Harris Creek, a good part of our watershed, and a lot of upland that seemed to have possibilities. Muriel and I put in an offer and we accepted the counter-offer. Enormously pleased, we transferred title to our children, Douglas, Colin, and Rachel.

As the price for the rest of the ranch approached market values, potential buyers began making serious enquiries. Muriel and I stood by, feeling that the family had enough land, and hoping that the eventual purchaser would be congenial and cooperative. But family opinions changed. Over Christmas 2016, Colin and Janet, his wife, said they would very much like to have their own place on the ranch. Both Douglas and Colin said they would rather have the ranch than an equivalent financial inheritance. The older grandchildren, now at university, declared that, whatever their futures, the ranch would figure in them. This, plus the fact that the price of the two lots that comprised the rest of the ranch was now lower than their appraised value, put a different cast on things.

Muriel and I talked the matter over, and my cautious and relatively business-wise wife opted strongly for purchase, the two of us sharing the cost. So did I, with almost equal enthusiasm. By this time there was another offer on the table, and we had to move quickly. An offer went in, and again we accepted a counter-offer. With the exception of my sister's land around the cabin, the Bosun Ranch now belongs to our children.

This has only just happened; I am still somewhat numb, and well aware of our excessive good fortune. The question hanging over the ranch throughout my life—its ownership through the next generation and beyond—has been settled. The ranch will stay in the Harris family for a good time yet. It could easily have been otherwise. The ranch was almost sold to the government in 1942, could easily have been subdivided after my grandfather's death in 1951, and things would have ended up very differently if John and Nancy had had children. John would have been horrified by this outcome, but, given the possible alternatives, perhaps Sandy, Mollie, and Nancy would have accepted it. Yet the ranch has come to us; by British Columbian standards it has a long and fascinating past, and it needs to be looked after, as for many years it hardly has been. Although a site of somewhat compromised privilege, contrary to what my grandfather told his Scottish bride it has never been an estate, rather a hillside farm that did not work very well. Now it is passing to another generation of the family. My children are taking over, but as an elder, I am not above musing a little about the meaning of this land, and offering a few thoughts about the future of a place that for as long as I can remember has been the geographical centre of my life.

By any economic calculus, the ranch has been an ongoing failure. Fruit from the orchard my grandfather planted did not have a market. For all the money and work that went into it, the ranch quickly became little more than a subsistence farm; eventually it was hardly more than a strip of cleared land on a mountainside. As his business family in Calne had thought, my grandfather was no businessman. He made decisions quickly, then poured his considerable energy into them—as when he abruptly took on the Slocan, cleared land, and planted a thousand-tree orchard. Shielded by temperament and an English inheritance, impulse overrode economic planning. Even in its prime, the ranch consumed money and work while yielding little product. Moreover, the political project to which my grandfather gave so much of his life yielded none of its anticipated results. The state did not take over the economy and manage it with expert efficiency. Individuals were not "conscripted for peace" and assigned to useful public service wherever. "Useless" people were not denied access to public services. In retrospect, his political agenda seems unworkable, even dangerous, and it is just as well that it also failed. To be sure, the early Slocan was full of hasty speculations that went awry, and the English Fabian Society, of which he was a member, was given to charming political dreams that, in retrospect, seem unfeasible. My grandfather was not alone on either count.

Yet there was nothing wrong with his commitment to sharing and cooperation, his strong sense of responsibility for others, his conviction, derived from his deeply held Christianity, that in giving we receive. There was nothing wrong, either, with his basic kindness. It brought him friends like the Bosun and Mrs. Hoshino. It made him very generous, especially with people he liked. I remember tagging along when two men came to buy some old farm equipment, which, although by this time he had very little money, my grandfather gave to them when he found they were Doukhobors. For all his grumbles about our side of the family, Sandy was also deeply kind, as both the hippies and the wild creatures of the ranch knew. My father too. When his memory was gone, his kindness remained. Whatever else, the ranch produced well-lived lives.

More generally, these lives were situated in circumstances that had been repeated over and over again across the span of Canadian settlement. Immigrants had left one land, crossed an ocean, and established themselves in another where circumstances were very different from those left behind. This was a recipe for cultural change, which began as soon as French settlers took up land along the lower St. Lawrence and around the Bay of Fundy. In the case of the ranch, life shifted away from its British roots. The cultural distance between Sandy and his father was large, and the basis of the difference was that their formative years had been lived, on the one hand in the midst of English society, and on the other on a mountainside farm in British Columbia. Moreover, the ranch was beginning to be situated in Canada. When my father went to university, he was sent not to an American or British school but to McGill in Montreal. Canada was coming into focus, and as it did, lives that were no longer quite British were being identified as Canadian. Long before the end of his life, even my grandfather, English as in many ways he remained, began to call himself a Canadian. His children, obviously no longer English, had no doubt about it.

And the ranch has been but an extreme example of the bounded patches of arable land stretched across the northern agricultural margins of Canada. In much of eastern Canada these patches pressed into the Canadian Shield, there to encounter acid soils, narrow growing seasons, and blackflies. Most of those taking on such land were poor, and their achievements meagre: relentless work, poverty, and the departure, sooner or later, of the young. The land was too hard, as was the struggle to rework it into viable farms. The more experienced, who often had come only for the timber, usually left quickly.

The ranch fits this pattern, but with the great difference that it caught an edge of the upper-middle-class English immigration to British Columbia at the end of the nineteenth century. There was money at the ranch, at least for a time, and some of the ways that accompanied it. But there was also a sharply bounded patch of land with limited agricultural capacity from which, inevitably, most of the ranch children would leave.

In short, the ranch is a complex inheritance to turn over to one's children. It delivers no single message, rather challenges that, in one way or another, have been wrestled with for years. Muriel and I trust our children; they will look after the ranch thoughtfully and creatively, and may even be willing to consider a few final words from their father.

It seems to me that the family should strive to put the ranch back into agricultural production, and probably the simplest way to do so, as my grandfather had concluded by the 1930s (see chapter 6) and Sandy many years later (see chapter 10), is to turn most of it over to livestock and hay meadow. To this end, the Far Field could be re-fenced, cattle pastured there, and the rest of the ranch fields upgraded to increase hay yields. This seems the most immediately feasible solution, and the ranch itself provides evidence that it can work. A much more daring solution would be to turn the ranch into a mixed teaching farm. The opposite of mechanized monocultures, it would include a variety of livestock, grains, small fruits, a diverse orchard, meadow, pasture, and gardens. People would come to learn how to farm the small, out-of-the-way patches of arable land that, like the Bosun Ranch, are scattered along the agricultural margins of Canada. Such land cannot accommodate monocultures and their equipment, yet has the potential to produce needed food for local markets. Perhaps, in changing times, the type of thoroughly mixed farm of which my grandfather so much approved is becoming feasible. But such farms, especially with a teaching component, are an enormous amount of work. None of my children has the time or skills, but many in the community do. The impetus would have to come from there, and someone would have to take charge.

In one way or another, the Bosun Ranch has been part of a larger community. In the early days, people frequently came up for tea. A few came for political discussion. Later, people came to visit Sandy, more of them, often, than he wished. Kids came to fish in Bosun Lake, and when, with a twinkle in his eye, Sandy caught them there he was kind and welcoming. Such encounters are remembered, part of growing up in the Slocan. In our day, people from the community are often around, most of them friends, and the ranch responds with pleasure. It is not a park, but nor is it a family retreat. It is a somewhat detached part of a larger community, and I hope this relationship is maintained and enriched.

Especially after my mother arrived, the ranch has attracted artists, and there has always been much family singing, music-making, and play-reading. In our day, Rachel's dance concert on the lawn in front of the ranch house has become an annual event. She brings excerpts from her current repertoire of contemporary Montreal dance, and creates improvisations that have turned our corner of the ranch into a huge stage and drawn her audience in her train. Artists of many stripes would love to be on the ranch. There could be artistic retreats, workshops, concerts for the community. The possibilities are enormous, but so, after a certain point, are the work and the costs. I only urge that the artistic connection be kept in mind.

For most of the last 120 years, the ranch has been a site of social criticism. The world at large has been seen to be too materialistic, too greedy, too urban, and far too unequal. Moreover, the whole socio-economic system has been seen to be tottering, its demise almost at hand. For some, the ranch has been a haven. This critique of society at large ran through all my grandfather's writing, as well as the rhetoric of hippie protest, and the gentler anarchism of recent back-to-the-landers. Their prescriptions for change, however, were very different. The hippies would have had no use for my grandfather's vision of a society managed by government experts, nor he for hippie ways stemming, he would have thought, from vastly exaggerated notions of individual freedom. And, although he would have had more sympathy for the anarchists, he thought that active government, rather than mutual aid in local societies, was required to replace the existing order. I doubt that either my grandfather or the hippies had a workable agenda for social change. Even the anarchist vision, however attractive locally, inadequately addresses the world's most pressing problems: poverty, the growing gulf between rich and poor, and the world's rapidly warming climate. Yet it seems clear that criticism in our deeply stressed world is essential, that the ranch's critical tradition should continue, and that the use of the ranch itself can be part of the critical process. Times are different, and criticism will be rethought, but in light both of the ranch's past and current world circumstances a complacent ranch would seem feckless and wasted.

Such are my more tangible musings as Muriel and I make this purchase and transfer the land we have bought to our children. Less tangibly, I am reminded of many a late summer afternoon when, walking alone in a ranch field, I have been awed by the beauty around me and by the simple fact that I was alive in this extraordinary place. The land glistened and life seemed wonderful and infinitely precious, a gift beyond all others. The gift of God perhaps, but I would rather say the gift of the last three billion years of life on earth. The ranch, as it is today, is a point in a huge reach of time. And, looking outward to and beyond the stars, it is situated in an unimaginably huge reach of space. The ranch sits, miraculously, at a particular intersection of time and space. So, at a slightly different scale, does the rest of our earth. Both share the same miracle. Presumably other miracles are scattered through the universe, but they won't be the same as ours, and we are its offspring and custodians. I feel this about the earth, and also about the ranch. When the Indigenous peoples of Canada invoke "Mother Earth," I am with them.

Muriel and I are leaving our children a site of wonder, and the opportunity to engage with it in creative ways that, now, can hardly be imagined.

ACKNOWLEDGMENTS

THIS BOOK IS EMBEDDED IN SEVERAL FAMILY GENERATIONS, AND COULD NOT have been written without their contributions. In his political booklets and dozens of letters to the editor of the *Nelson Daily News*, my grandfather left a broad trail to his political thought (chapter 6), but it was my mother who encouraged him to write about his coming to Canada. During the last years of his life he did, often repeating himself, but leaving the dozens of typed, single-spaced foolscap pages that are the basis of the first two chapters of this book as well as of a good part of chapter 4. Notes from conversations years ago with my father and Uncle Sandy have allowed me to write most of what I have about the ranch in its early, prime years (chapter 3). The discovery of my father's diary, written in 1924 while he was working on the ranch and finishing a thesis for McGill on Bernard Shaw, allowed me to describe a failing ranch economy, something of the ranch's political soul, and decisions about who in the second generation would stay and who would leave (chapter 5). My mother's album of photographs of the log cabin near Loch Colin, which my father largely built but she inspired, has enabled me to describe the making of that beautiful building (chapter 7), which became the ranch anchor of the R. C. Harris side of the family.

Most of the basic legal documents, as well as the correspondence associated with my grandfather's death in 1951, have survived, and are the basis for the short chapter on his will and its contentious aftermath (chapter 9). If Sandy had not written to his daughter, Nancy, each week for almost thirty years, if she had not kept his letters, and if, long after her death, her foster son, Ralph Wilson, had not passed this voluminous correspondence on to me, it would have been impossible to write more than a brief, anecdotal account of Sandy's many years at the heart of the ranch (chapter 10).

The chapters on restoring the ranch house (chapter 11) and building the clay house (chapter 13) depend on recent memories—Muriel's and mine, as well as those of the people who worked on these buildings. I have kept in touch with most of the young Americans who rebuilt the ranch house, and in the last couple of years have had many good conversations with them about those intense, early 1970s years when the ranch house was largely rebuilt. The clay house, on the other hand, is not yet ten years old, most of the people who built it are our good friends, and the account of its building, while largely mine, bears their imprint, as well as Muriel's and our children's. The chapter on the aftermath of the second generation (chapter 12) has generated a discussion, reaching to the eldest of the fifth generation, about the pictures that best represent our family's recent use of the ranch.

In short, most of this book is embedded in family writing spread over more than a century, in a few family stories, and in a considerable accumulation of recent memories.

However chapter 8, on the interactions of the Harris family with Japanese Canadians during their years on the ranch, reaches well beyond such sources. Mine is a local account, but the treatment of Japanese Canadians during the war remains the subject of intense discussion in Canada, and even a local account is drawn into its orbit. This chapter is a small part of that discussion, and in this realm, where I am a neophyte, I have relied on many others: English cousins Walford Gillison and Frances Clemmow, who provided information about members of the Harris family's missionary work in China and internment there by the Japanese during World War II. Momoko Ito, manager of the Nikkei Internment Memorial Centre in New Denver, who made the museum's collections available to me and offered important advice, as did Beth Carter and Linda Reid at the Nikkei National Museum. Frank Moritsugu, journalist and author of a book on Japanese teachers in the camps, who reminded me of my grandfather's writings on the Japanese Canadians in the British Columbia Archives. Slocan friends Tsuneko Kokubo, Paul Gibbons, and Taeko Miwa, who commented on a draft of the manuscript. Academic friends Jordan Stanger-Ross, Peter Ward, and Peter Ennals, who did the same. Within the family, my children and a niece offered particularly useful criticisms.

My relations with Harbour Publishing have been cordial and efficient throughout. Pam Robertson, an exemplary editor and a pleasure to work with, tweaked the manuscript as only a fine editor can. Nicola Goshulak has been a meticulous copy editor. I have much appreciated the talents and efficiency of the crew at Harbour Publishing that guided the book through its late stages.

At a late stage, various members of the family, including my sister and her husband, helped locate and assemble the pictures and offered final comments on the text. Above all, my wife, Muriel, kept details in mind and the momentum going when, grounded by radiation treatments for a large abdominal tumour and then by a major surgery, I had almost no energy to do so. She has my deep appreciation and boundless affection.

Cole Harris
Vancouver, BC
November 2017

NOTES

CHAPTER 1 CALNE TO COWICHAN

1 Harriet Cooper, "History of the Harris Family at Calne, 1775 to 1907," typescript, Harris family papers, p. 23.

2 John, Tom, and Henry Harris were all children by Thomas Harris's second wife, Sophia Mitchell, who died in 1864. The second bed seems to have begot business people, the third, with Elizabeth Colebrook, doctors, church people, and impractical intellectuals.

3 I have found little information about my great-grandmother, Elizabeth Colebrook. These fragments about her family come from *The Family of William Colebrook, 1801–1869*, a book prepared by Gladys Stanford and Peter Layng, printed by the Courtney Press Ltd., Castle Street, Brighton, and circulated among the extended Colebrook family.

4 Harriet Cooper, "History of the Harris Family," p. 27.

5 Unless otherwise indicated, all the quotes in this chapter attributed to J. C. Harris are taken from the recollections of his early years in Canada that, over a period of years from December 1943 to February 1951, he wrote in response to a request from my mother. All are in the J. C. Harris fonds, Royal British Columbia Archives, Victoria. This quote is from J. C. Harris to Ellen Harris, Dec. 1, 1943, JCH fonds, RBCA.

6 Harriet Cooper, "History of the Harris Family," p. 38.

7 J. C. Harris to Ellen Harris, Dec. 1, 1943, JCH fonds, RBCA.

8 After graduating from Mill Hill, he played briefly for the Harlequin Rugby Football Club, a North London club, including in a game against the French national side, which the Harlequins won.

9 J. C. Harris to Ellen Harris, Dec. 1, 1943, JCH fonds, RBCA.

10 "Fulford Harbour," June 25, 1945, JCH fonds, RBCA.

11 "Saltspring Island," JCH fonds, RBCA. Captain Barkley was the grandson of Captain Charles William Barkley, after whom Barkley Sound, on the west coast of Vancouver Island, is named. See J. F. Bosher, *Imperial Vancouver Island: Who was Who, 1850–1900*, self-published, 2010, pp. 111–12.

12 "Saltspring Island," JCH fonds, RBCA.

13 "Life at Hall's Crossing," JCH fonds, RBCA.

14 *Ibid.*

15 Alexander C. Harris and Joseph C. Harris, "Our Diary in America from June 17–Oct. 1,1891," RBCA, pp. 49–50.

16 "Life at Hall's Crossing," JCH fonds, RBCA.

17 *Ibid.*

18 *Ibid.*

19 "Cowichan Remembrances," JCH fonds, RBCA.

20 "None of us," Joe wrote later, "could see any sense in Bess and Mary going out to China as missionaries, and thought that they were throwing their lives way." He, Willie, and Alec were of this opinion, as was their father.

21 "Cowichan Remembrances," JCH fonds, RBCA.

22 "Last Years in Cowichan," JCH fonds, RBCA.

23 *Ibid.*

24 *Ibid.*

25 *Ibid.*

CHAPTER 2 PROSPECTING FOR LAND

1 For a longer version that includes comments about a later Slocan, accounts of his off-farm work, and longer descriptions of particular individuals, see Joseph Colebrook Harris, *Beginnings of the Bosun Ranch*, booklet no. 2, Slocan History Series, New Denver, Chameleon Fire Editions, 2015.

2 The longer but much easier route was down the Slocan River to the junction with the Little Slocan, then up it to the meadows in question. The route taken required bush-whacking through mountains.

3 Joe's description of this sleigh road: "Of course the sleigh road that we managed to build with the first effort [spring of 1897] was very poor. We chopped the trees to fall along the roadway on the lower side, and then as much dirt as possible was shovelled onto the brush pile. It made a quite passable road for the first winter as long as the dirt was frozen, but it was only the beginning of a real road; the next winter when I did a great deal of heavy hauling on it (ice and wood) it was in far worse condition, and the outside had sunk down and on several occasions, the sleigh went over the bank, luckily without much damage. It was often a terrific job to pass another sleigh."

4 This log building, quickly built in the fall of 1897, remains and has been restored. See chapter 11.

5 The market problems, which would always bedevil the Bosun Ranch, were emerging as early as 1898.

CHAPTER 3 THE EXPANSIVE YEARS: 1898–1918

1 A decade later he told the Farmers' Institute in Nakusp that the first step in planting an orchard should be to get married. For a man to "plant an orchard without a wife to help and advise him, and some boys and girls to work for... [was] an absurdity. This was the first mistake I made and it was a very serious one, but I soon saw the error of my ways..." J. C. Harris, "Common Mistakes Made in Setting Out Orchards by One Who

Has Made Them," Farmers' Institute Report, *Sessional Papers of British Columbia*, 1909, p. K 26.

2 According to my grandfather, the problem would be lessened were the government to provide telephones to all farm houses. J. C. Harris, "Common Mistakes," p. K 27.

3 Ellen Harris to Dick Harris, Bosun Ranch, September 1941, Harris family papers.

CHAPTER 4 THE BOSUN MINE

1 My grandfather's writings about the Bosun Mine are published in J. C. Harris, *Beginnings of the Bosun Ranch*, booklet no. 2, Slocan History Series, Chameleon Fire Editions, New Denver, 2015.

2 Wildcats were claims, acquired as pure speculations, on which no exploratory work had been done.

3 According to an article in the *British Columbia Mining Record* (vol. 5, no. 8, Aug. 1899), Delayney would have had trouble substantiating his right to these claims had it been called into question. "Peace at any price, however, appears to be the motto of the rancher, and so having disposed of likely complications he proceeded with his task of cultivating the soil, a wiser though apparently poorer man."

4 Prospectors were frequently "grubstaked" by local merchants who supplied provisions in return for a half share in any discoveries.

5 According to *The Paystreak* (Sandon, July 28, 1897), "The strike consists of a six or eight inch strata of solid galena running at 160 odd ounces of silver and 77% lead."

6 The *British Columbia Mining Record* (vol. 5, no. 8, Aug. 1899) provided additional details: "He [W. H. Sandiford] agreed to keep two men continuously at work on the property for 30 days, at the expiration of which he had the privilege of either relinquishing all title, paying 10 per cent down on a year's bond at $15,000, or purchasing at a cash consideration of $7,500." Forty feet down, there was ore across almost the full width of an experimental shaft. "It is not to be wondered at that Mr. Sandiford accepted the latter alternative in the terms of the agreement, and Mr. Harris, the owner of the ranch, who had been deluded into paying $700 for two worthless claims, found himself the richer by $7,500."

7. *British Columbia Mining Record*, vol. 5, no. 8, Aug. 1899.

8 The building that would later become the log core of the ranch house.

9 Arthur Cleverley, a lad from Calne, was still at the ranch with my grandfather at this point.

10 *British Columbia Mining Record*, vol. 5, no. 8, Aug. 1899.

11 Charlie must have arrived in the summer of 1899. He appears in the 1901 Canadian census, then twenty-one years of age and identified as a mining engineer.

12 *British Columbia Mining Record*, vol. 9, no. 10, Oct. 1902.

13 *British Columbia Mining Record*, vol. 12, no. 9, Sept. 1905.

14 *Ibid.*

15 *British Columbia Mining Record*, vol. 13, no. 9, Sept. 1906.

16 Report, Minister of Mines, 1929, *British Columbia Sessional Papers*, 1931, vol. 1, section C, p. 316.

17 Sandy Harris to Nancy and John Anderson, Oct. 20, 1956, Harris family papers.
18 Richard White, *Railroaded: The Transcontinentals and the Making of Modern America*, New York, W. W. Norton, 2011.

CHAPTER 5 THE BOSUN RANCH, 1924

1 This picture of the ranch economy is amply borne out by the ranch account books, which have surfaced since I wrote this section. They show a great variety of sales (fruit, vegetables, meat, eggs, milk, butter, wood, ice), none of which was very large. A quarter to a third were not paid for. The monthly return after expenses was in the vicinity of one hundred dollars. Expenses did not include family labour, of which the ranch consumed huge quantities.
2 Joe thought that the local Farmers' Institute should investigate the possibility of hiring an agent to market Slocan fruit, but nothing came of the idea, probably because the Slocan's fruit production was too small to warrant the expense. J. C. Harris, "A Talk on the Work of Farmers' Institutes," Sessional Papers of British Columbia, 1911, pp. L 152–53.

CHAPTER 6 J. C. HARRIS, SOCIALISM, AND THE FABIAN IDEAL

1 Dick Harris to Ellen Code, Vancouver, March 7, 1930, Harris family papers.
2 R. C. Harris, draft of an account of his father and of the ranch, p. 7, Harris family papers.
3 My grandfather claimed that he could get two hay crops a year without irrigation, and three or four with irrigation and silage. He had added a silo adjacent to the main barn. J. C. Harris, "Alfalfa Possibilities in West Kootenay," unpublished typescript, ca. 1940, Harris family papers.
4 J. C. Harris, "Alfafa Possibilities," p. 2.
5 Dick Harris to Ellen Harris, Aug. 29, 1930, UBC Special Collections, Ellen Harris fonds, Box 41, file 7; Ellen Harris to Dick Harris, Aug. 3, 1932, UBC Special Collections, Box 39, file 5.
6 J. C. Harris, *British Columbian Problems*, Thomas Stationery Co., Vancouver, BC, 1909.
7 J. C. Harris, *British Columbian Problems*, p. 30.
8 Henry Drummond, *Natural Law in the Spiritual World*, New York, William L. Allison Company, 1889.
9 At the very end of his life, he acknowledged Drummond's influence and sketched his religious beliefs in a letter to a former Sunday School pupil who was questioning her own faith. J. C. Harris to Emma Clever, New Denver, March 27, 1951, Harris family papers.
10 See *Natural Law in the Spiritual World*, particularly the preface and introduction.
11 J. C. Harris, *Conscription for Peace*, Harris fonds, RBCA; also Harris family papers, p. 16.
12 This phrase is from a letter to the *Nelson Daily News*, submitted Jan. 20, 1932, Harris family papers.
13 In 1946 alone, twenty-six letters from my grandfather appeared in the *Nelson Daily News*.

14 "Kerr Called Down," *The Prospector*, Kaslo, July 25, 1895.

CHAPTER 7 THE LOG CABIN

1 Ellen Harris to Dick Harris, Bosun Ranch, Sept. 20, 1941, Harris family papers.

2 *Ibid.*

CHAPTER 8 THE JAPANESE CANADIANS

1 Frances Clemmow, *Days of Sorrow; Times of Joy. The Story of a Victorian Family and Its Love Affair with China*, Kibworth Beauchamp, Leicestershire, Matador Publishing, 2012, p. 166.

2 J. C. Harris to progeny, New Denver, May 27, 1942, Harris family papers. Sandy, his son, and Mollie, Sandy's wife, had been spending their winters at the ranch house.

3 *Ibid.*

4 Margaret Harris to Ellen Harris, undated, but included in a letter from J. C. Harris to progeny, May 27, 1942, Harris family papers.

5 A. C. Taylor to E. L. Boultbee, Aug. 14, 1942, NAC, RG 36/27, box 15, file 1020, 1942. Also Department of Labour, Japanese Division, NAC, RG, 36/37, box 26, file 1020, 1946. The "north forty acres" is the Far Field, where the Japanese houses were located.

6 R. C. Harris, "Joseph Colebrook Harris," handwritten draft manuscript, no date, Harris family papers.

7 J. C. Harris, "Notes and Memories of the Coming of the Japanese to the Slocan Lake Country, 1942," written in late 1944 or early 1945, RBCA, Add. MSS 807 Box 3, file 3. The quotations and interpretations in this paragraph are all from this source.

8 Here J. C. Harris refers to an incident in 1914 when most of the 376 passengers—all of them British subjects and the large majority Sikhs from India—on a Japanese ship, the *Komagata Maru*, were detained in Vancouver harbour and denied entry to Canada.

9 Here J. C. Harris underestimated the pervasive reach, much wider than the working class, of racism throughout the late nineteenth- and early twentieth-century Anglo-American world.

10 Anti-Japanese sentiment, further whetted by news of Japanese atrocities in Hong Kong and the Philippines, was as virulent as my grandfather suggests, even on the political left. See Werner Cohn, "The Persecution of Japanese Canadians and the Political Left in British Columbia, December 1941–March 1942," *BC Studies* 68, winter 1985–86, pp. 3–22.

11 See also J. C. Harris, "Our New Canadians' Problems," no date but probably 1945, RBCA, Add. MSS 807, box, 3, file 12.

12 The quotations and interpretations in this paragraph are all derived from Harris, "Notes and Memories," and "Our New Canadians' Problems."

13 J. C. Harris, "Notes and Memories." The other quotation in this paragraph is also from this source.

14 Patricia E. Roy, "If the Cedars Could Speak: Japanese and Caucasians Meet at New Denver," *BC Studies* 131, autumn 2001, p. 84.

15 Kaslo, like New Denver, had accepted Japanese evacuees, and Japanese community relations probably unfolded there much as they did in New Denver. See Patricia E. Roy, "A Tale of Two Cities: The Reception of Japanese Evacuees in Kelowna and Kaslo, BC," *BC Studies* 87, autumn 1990, pp. 23–47.

16 Frank Moritsugu, *Teaching in Canadian Exile: A History of the Schools for Japanese Children in British Columbia Detention Camps during the Second World War*, Toronto, Ghost-Town Teachers Historical Society, 2001, p. 73.

17 Sandon Hotel Old Men to Mr. George Collins, Commissioner, BC Securities Commission, Sandon, May 9, 1944, Sandon Museum Collection.

18 I have not found a reliable figure for the number of evacuees on the Far Field. In the Orchard in New Denver in 1945 there were 1,277 Japanese Canadians in 200 houses, an average of 6.4 people per house. In the camps in and around New Denver at the end of 1942 there were 1,505 evacuees in 293 houses, an average of 5.2 people per house. At 6 people per house there would have been 150 people on the Far Field.

19 J. C. Harris, "Notes and Memories."

20 Margaret Raeper Harris to Ellen Harris, Bosun Ranch, July 5, 1943, Harris family papers. Uncle Sandy said that the garden went downhill until the Japanese Canadians came. "Then it was so beautiful it made you gasp." Notes of a conversation with Sandy, Aug. 1985, Harris family papers.

21 In the summer of 1943 my grandmother, debilitated by Parkinson's disease, was near the end of her life. On Christmas Eve, 1943, my grandfather wrote to tell my parents that "Granny" was in the hospital and very weak. "My dears it is awfully sad, but we have had such wonderful times and been so happy and united. I do not think that we can wish for her suffering and weakness to continue long. We are in God's hands and He is very merciful." J. C. Harris to Ellen and Dick Harris, New Denver, Dec. 24, 1943, Harris family papers. "Granny" died a few weeks later.

22 Basil Izumi's memories of the ranch, where he lived from ages six to eight, may be fairly typical. For him there was a lot to do on the ranch, summer and winter, and life there was fun. Basil Izumi, personal communication.

23 Patricia E. Roy, "If the Cedars Could Speak: Japanese and Caucasians Meet at New Denver," *BC Studies* 131, autumn 2001, p. 87.

24 This information is from Sachi Manuel, a young teenager on the Bosun Ranch during the war, now living in Olympia, Washington.

25 Notes from a conversation with Sandy, Aug. 5, 1985, Harris family papers.

26 Sandy told these stories over and over, and particularly relished the symbolism of Tak's encounter with the Canadian flag. Tak was born in Canada, and Sandy considered him a Canadian through and through. My wife, Muriel, noted down these accounts in the old ranch house on Aug. 5, 1985.

27 Debates, House of Commons, 19th Parliament, Fifth Session, Vol. 6, Aug. 4, 1944, pp. 5915–17.

28 J. C. Harris to ed., *Nelson Daily News*, Sept. 3, 1943; cited in Roy, "If Cedars Could Speak," p. 89.

29 J. C. Harris, "Notes and Memories."

30 J. C. Harris to Mackenzie King, Feb. 23, 1946, NAC, King Papers, no. 365411.

31 J. C. Harris to Mrs. Hoshino, June 22, 1949, RBCA, Add. MSS 807, box 1, file 28.

32 For a detailed published account of the events described in this and the following two paragraphs, see the chapters on Yangchow in Greg Leck, *Captives of Empire: The Japanese Internment of Allied Civilians in China 1941–1945* (Shandy Press: SAN 257-0181, 2006). My remarks, however, are largely derived from the recollections of my second cousin Walford Gillison—the boy identified here—now living in Somerset, England.

33 I am indebted for most of the points in this paragraph to conversations and correspondence with Jordan Stanger-Ross, associate professor in the Department of History at the University of Victoria, and to a draft of his paper "Suspect Properties: The Decision to Dispossess Japanese Canadians," which he has kindly allowed me to read.

34 As late as 1947 the premier of British Columbia was urging the federal government to make the ban on "Japanese" permanent. For a good discussion of attitudes in BC and Ottawa toward resettlement and "repatriation," see Greg Robinson, *A Tragedy of Democracy: Japanese Confinement in North America* (New York: Columbia University Press, 2009), especially pp. 261–74.

35 Debates, House of Commons.

36 For an example of the racism experienced on Saltspring Island both before and after the war by a family that lived in the Rosebery Camp, see Rose Murakami, *Ganbaru: The Murakami Family of Salt Spring Island* (Saltspring: Japanese Garden Society of Salt Spring Island, 2005).

37 Quoted by Frank Moritsugu, *Teaching in Canadian Exile*, p. 220.

CHAPTER 9 DEATH AND SUCCESSION

1 Two days before, he ended a long letter about Darwin and God with the following: "My life is pretty precarious, as my old heart is worn out and may quit at any time. I also want new eyes, legs, and ears." J. C. Harris to Emma Clever, Victoria, March 27, 1951, Harris family papers.

2 Heather Rose to Sandy Harris (draft) and note to my father, Harris family papers.

3 Nancy to Heather (carbon copy), New Denver, June 19, 1959, Harris family papers.

4 Sandy acquired all of the Bosun Ranch (almost three hundred acres, most of them mountainside) except the land around my parents' cabin (some fifteen acres) and the parcel (some ten acres) belonging to the Bosun Mine.

5 Dick Harris to Sandy Harris, Vancouver, March 12, 1961, Harris family papers.

CHAPTER 10 SANDY'S RANCH

1 Sandy's letters are held by the family. Specific letters are referenced by date.

2 Draft of a letter from A. L. Harris to Mr. G. A. MacMillan, New Denver, Aug. 16, 1952, Harris family papers.

3 These people, among the key founding figures of the CCF, came to the West Kootenay because they thought it a winnable seat, and stayed at the ranch because the ranch

house provided a warm welcome, comfortable quarters, and no end of congenial political talk.

4 My account of Lilly's recollections of her time on the Bosun Ranch are based on two interviews with her, one on July 30, 2014, the other on Oct. 25, 2015, both at the old ranch house.

5 Gretchen's recollections of her time in the Ranch House are derived from two interviews with her in Vancouver, one in 2014, the other in 2016.

6 When Mollie arrived at the nursing home in Vernon, Sandy was asked to transfer her old age pension to the home in partial payment for her care. But neither Sandy nor Mollie had taken their pension on the grounds that they had enough to get by, and did not want to make unnecessary claims on the state. The matter was rectified; Mollie's pension was claimed and made over to the nursing home, and Sandy may finally have taken his pension as well.

7 The last ten years of Sandy's life are best remembered by Ralph Wilson, who arrived on the ranch as a foster child in 1975–76 and lived there for most of the next ten years. He reports that Sandy was always working in the garden, or walking with his dog, or feeding fish, and that he had no mean bone in his body. Interview with Ralph Wilson, Bosun Ranch, Oct. 25, 2015.

8 Rupert Brooke, *Letters from America*, London, 1916, chapter 13.

CHAPTER 11 THE COUNTERCULTURE AND THE RANCH HOUSE

1 Thomas now lives in Point Roberts, and much of this biographical information comes from conversations there with him in the spring of 2014.

2 Lilly, who now lives in Hawaii, returned to New Denver in the summer of 2014, and I talked with her then about her Slocan years.

3 Thomas Wright to Cole Harris, Sandon, Sept. 8, 1972, Harris family papers.

4 Thomas Carlyle, *Chartism*, London, 1840.

5 Sam Tichenor still lives in the Slocan, and in September 2014, we talked about his life and his work on the ranch.

6 Thomas Wright to Cole Harris, November 1973, Harris family papers.

7 Gretchen now lives in Vancouver where, in February 2015, I talked with her about her time in the old ranch house.

8 David Lehton now lives in North Saanich, Vancouver Island, and I talked with him there in March 2015 about his time on the ranch.

9 Bob McKay, now a wood turner, lives on Saltspring Island, and in February 2015, I talked with him there about his background, his Slocan years, and his work on the ranch house.

10 Glenn Jordan still lives in the Slocan. I have frequently talked with him about his past, most recently in September 2014.

INDEX

Page numbers in **bold** refer to photos and illustrations.